Planning and Conducting
Applied Agricultural Research

Also of Interest from Westview Press

† *Farming Systems Research and Development: Guidelines for Developing Countries,* W. W. Shaner, P. F. Philipp, and W. R. Schmehl

Readings in Farming Systems Research and Development, edited by W. W. Shaner, P. F. Philipp, and W. R. Schmehl

Farming Systems in the Nigerian Savanna: Research and Strategies for Development, David Norman, Emmy B. Simmons, and Henry M. Hays

XIV International Grassland Congress, edited by J. Allan Smith and Virgil W. Hays

* *Agricultural Economics: Principles and Policy,* Christopher Ritson

Agroclimate Information for Development: Reviving the Green Revolution, edited by David F. Cusack

† *Science, Agriculture, and the Politics of Research,* Lawrence Busch and William B. Lacy

Azolla as Green Manure: Use and Management in Crop Production, Thomas A. Lumpkin and Donald L. Plucknett

Wheat in the Third World, Haldore Hanson, Norman Borlaug, and R. Glenn Anderson

* *An Introduction to the Sociology of Rural Development,* Norman Long

† Available in hardcover and paperback.
* Available in paperback only.

About the Book and Authors

Planning and Conducting Applied Agricultural Research
Chris O. Andrew and Peter E. Hildebrand

This study focuses on applied research as a service to a client with a problem that research can help solve. Because applied research has a definite purpose, there is usually a time constraint, a deadline by which the work must be completed, as well as a limit on the resources the client has available or is willing to use. Consequently, the researcher must concentrate on the efficient use of the research resources while trying to maximize the likelihood of providing a useful product. Professors Andrew and Hildebrand offer an approach to identifying researchable problems and proceeding efficiently to their resolution. Their material can be used effectively both in teaching and by individuals working in the field.

Dr. Chris O. Andrew is professor in the Food and Resource Economics Department and associate director of the Center for Tropical Agriculture and International Programs at the Institute of Food and Agricultural Sciences (IFAS), University of Florida. **Dr. Peter E. Hildebrand** is professor in the Food and Resource Economics Department and coordinator of the Farming Systems Research and Extension Program at IFAS.

Planning and Conducting Applied Agricultural Research

Chris O. Andrew
and Peter E. Hildebrand

Routledge
Taylor & Francis Group

LONDON AND NEW YORK

First published 1982 by Westview Press

Published 2019 by Routledge
52 Vanderbilt Avenue, New York, NY 10017
2 Park Square, Milton Park, Abingdon, Oxon OX14 4RN

Routledge is an imprint of the Taylor & Francis Group, an informa business

Library of Congress Cataloging in Publication Data
Andrew, Chris O.
  Planning and conducting applied agricultural research.
  (A Westview special study)
  Includes bibliographical references.
  1. Agricultural research—Planning. 2. Agricultural research—Methodology.
I. Hildebrand, Peter E. II. Title. III. Title: Applied agricultural research.
S540.A2A53   1982   630'.72   82-13503

ISBN 13: 978-0-367-28296-7 (hbk)
ISBN 13: 978-0-367-29842-5 (pbk)

TABLE OF CONTENTS

PREFACE ix
ACKNOWLEDGMENTS xi

CHAPTER I:
INTRODUCTION 1
    Applied Research 3
    The Book 4

PART ONE: PLANNING APPLIED RESEARCH

CHAPTER II:
EFFECTS OF RESOURCE AVAILABILITY ON APPLIED
    RESEARCH 6
    Information Resources 7
        Secondary and Primary Information 8
        Time Series and Cross-Section Data 9
        Experimental and Non-Experimental Data 9
    Human Resources 10
    Physical Resources 11
    Financial Resources 12
    Time Constraints 12
    Summary 13

CHAPTER III:
ORIENTATION AND FOCUS OF PROJECTS: RESEARCH-
ABLE PROBLEMS, HYPOTHESES,
    AND OBJECTIVES 14
    A Conceptual Model 14
    Specification of a Researchable Problem 16
        Problems Reflect Felt Needs 17
        Problems are Non-hypothetical 17
        Problems Suggest Meaningful, Testable
            Hypotheses 18
        Problems are Relevant and Manageable 19
        Researchable Problems vs. Problematic
            Situations 19
        Examples of Problem Statements 20
    Formulation of the Hypotheses 23
        Characteristics of Hypotheses 24
        Some Examples of Hypotheses 24
    Delineation of the Objectives 28
    Summary 30

PART TWO: CONDUCTING APPLIED RESEARCH

CHAPTER IV:
  EXPERIMENTAL DATA COLLECTION 34
     Experimental Design 35
        Relationship to the Problem 35
        Relationship to Resources.................... 38
     Secondary Experimental Data................. 41
     Multi-purpose Experimentation 42
     Multi-disciplinary Experimentation 45
     Summary......................... 46

CHAPTER V:
  NON-EXPERIMENTAL DATA COLLECTION 47
     Selecting Respondents 47
     Designing the Questionnaire 49
        Difficulties in Interpretation and
           Communication 50
        Designing for Data Retrieval 51
     Pretesting the Questionnaire 52
        Size of Pretest 53
        Information Checking 54
        Time Difficulties 55
     Selecting and Training Interviewers 56
     Verifying Primary Data 57
     Verifying and Using Secondary Data 58
     Summary......................... 60

CHAPTER VI:
  DATA UTILIZATION — WHAT DOES IT ALL MEAN? .. 61
     Flexibility of Interpretation...................... 63
        Meaning of the Results 63
        Reliability of Results...................... 66
     Presentation of the Results 69
     Summary......................... 71
REFERENCES 73
APPENDIX 79
INDEX 87

PREFACE

This book is the culmination of a group effort to eliminate a deficiency made evident during the organization of a graduate course in research methodology at the UN-ICA Graduate School in Agricultural Sciences in Bogota, Colombia.[1] The deficiency resulted from the difficulties associated with organizing research projects oriented toward real world problems and formulated so that 1) the research can be completed within the available period of time, and 2) the results will be useful in helping to resolve the problem toward which the study is directed. It became obvious to a group of agricultural economists[2] working with the graduate program that the various standard approaches to the presentation of research methodology are not successful in helping students become efficient researchers consistently able to make meaningful contributions to the resolution of agricultural and related problems of their country.

Initially the efforts of Michael Steiner who was responsible for the methodology class and of James Driscoll, Chris Andrew, and Peter Hildebrand, who were helping in the development of the material, focused upon new means of presenting the requisites for successful research in a manner that would have real meaning and utility for the students. Thus, the core idea now presented in Chapter III germinated. The topics are not new; most research methodology material discusses research problems, hypotheses, and objectives. It is the means of fitting these parts together into a more useful form which is new. During several courses and while counseling students and fellow staff members of ICA in research, we were able to revise and improve the concepts. We found at the same time that this approach to planning and executing applied research could be easily understood and used. As the success of the approach became clearer, we decided that it was worth the time required to present it in book form for a wider audience.

In the process of developing the book, to which all four of us initially contributed, we found that, although it was relatively easy to use the approach in training students and counseling researchers, it was difficult to present the approach in a form understandable and usable by persons with whom we would have no direct contact.

---

[1] The graduate school is jointly administered by the National University of Colombia (UN) and the Colombian Agricultural Institute (ICA). Besides the graduate school, ICA has responsibilities in research and extension as well as service activities such as control of agricultural chemicals and port sanitation.

[2] This group consisted of several Colombian agricultural economists with ICA including those mentioned in the acknowledgements and agricultural economists with the University of Nebraska Mission in Colombia. This technical assistance team worked with ICA, National University and the Graduate School from 1967 to 1972.

Because of our various commitments and spatial separation it became increasingly difficult for all four of us to coordinate our efforts, so finally Andrew and Hildebrand assumed the responsibility for the long process required to convert the vague ideas and concepts into a form which could be readily conveyed to students, researchers, and others.

The basic theme of this book is that of applied research as a service to a client with a problem for which the information obtained by research can help resolve. Because applied research has a definite purpose, there is usually a time constraint or deadline within which the work must be completed as well as a limit on the other resources the client has available or is willing to use in the resolution of the particular problem. Consequently, the researcher must be cognizant of the efficient use of research resources while at the same time functioning so as to maximize the likelihood of providing a useful product to the client.

Besides the approach to applied research (covered in the book), another important factor affecting the success a researcher will experience in serving his clients is the research environment within which he labors. An applied researcher cannot be effective in satisfying clients when he is isolated from them by a system that reduces or prevents effective communication between them.[3] This can happen, for example, when an extension service with direct client contact has little communication with research personnel even though they may be in the same organization. It can also happen in research organizations in which projects are dictated by administrators who have little contact with the clients and hence have no appreciation of their real problems. This may be an argument for maintaining only small research organizations, but more realistically, it is an argument for an organization in which the researcher maintains close personal contact with the clients and where he in turn shares in determining the research priorities of the organization. We suggest that a better coordinated working relationship between research administrators, researchers, and clients will develop if all three groups understand the approach to problem identification which is presented in this book.

Although a majority of the examples are authentic, and are drawn from Colombia where most of the writing was done, the research difficulties presented are not unique to that country; they are encountered throughout the world including the United States. Our desire is to provide the researcher, wherever he may be, with an approach to research under various time and resource limitations

---

[3]See [2, 30, 38, 60, 61, 65, 69, 75, 80, 82, 86, 87, 88] for readings on the role and impact of institutions in the research process.

which will help him be of greater service to his clients. This is particularly true in developing countries where applied research is so needed.

ACKNOWLEDGMENTS

To acknowledge the assistance of each individual for comments presented during numerous discussions and on the seemingly endless series of drafts leading to this book is impossible. Without the early dedication and ideas and comments by Mike Steiner (presently with the Armour Food Company) and Jim Driscoll (presently with the Economic Research Service of the United States Department of Agriculture) this book would not have been initiated nor, possibly, completed. Two Colombians, Juan Acosta and Ramiro Orosco, at the Colombian Agricultural Institute (ICA) deserve special recognition for reviews of the material and constructive criticism, and for using concepts and early drafts in the classroom. Likewise to Rafael Samper, Department Head in Agricultural Economics at ICA, and his staff, gratitude is due for sustained interest and encouragement.

Ideas and most of the drafts for the book were forthcoming while the authors were under contract with the University of Nebraska at the Colombian Agricultural Institute and the National University of Colombia. This contract, funded by the Ford Foundation and the USAID, helped develop the agricultural economics profession in Colombia including three undergraduate programs, a graduate program, and both research and extension programs. To our colleagues, and to students at these institutions, we offer sincere acknowledgments and hope that the book will be useful to them.

We express our appreciation for assistance received from the Food and Resource Economics Department at the University of Florida. To Fred Prochaska and his students who have used the text for two years in the research methodology course, we are grateful for constructive criticism. Special recognition is due to Leo Polopolus, Chairman of the Food and Resource Economics Department, and W. W. McPherson, Graduate Research Professor, for reviews and consultation. Special appreciation is extended to Beth Davis for supervising preparation of the final manuscript for printing.

Also, we extend our appreciation to the Ministry of Agriculture of El Salvador, where the second author was stationed on a technical assistance contract, for translation and preliminary publication of the manuscript in Spanish. For final reviews of the Spanish translation and preparation of the final manuscript for printing, gratitude is extended to the Guatemalan Institute of Agricultural Science and Technology where the second author was employed from 1974 to 1979.

Lastly, we want to express our sincere gratitude to our families who shared their time for numerous weekends and nights with an idea which may have seemed vague and boring.

Chris O. Andrew

Peter E. Hildebrand

Planning and Conducting
Applied Agricultural Research

CHAPTER I

INTRODUCTION

About midmorning, the Minister of Agriculture is just completing a phone conversation as the door to his office opens to admit one of the young men in the research group from the Planning Department of the Ministry. At about the same time the Vice Minister and the head of planning enter and the four men seat themselves around a small conference table. The Minister opens the meeting, directing his comments to his Vice Minister: "I'm told that we have a final report on that project on importing fertilizer, that's great. It's just in time. Now we ought to be able to convince the import-export people that they can't reduce our request for fertilizer and still expect us to meet our new trade commitments."

Turning to the young researcher, the Minister says encouragingly, "OK, young man, they tell me you've made an excellent study. Now tell us exactly how much fertilizer is it going to take over the next five years to meet our production targets?"

"Sir?" replies the researcher, apparently a little confused.

"Come, come, now, don't be nervous. Just tell us about the research you've been doing. What are the results?" The Minister realizes the researcher may be a bit timid.

"Oh, yes sir," smiles the researcher, "thank you."

"Well, as soon as we got your request for information on the importance of fertilizer, we checked to see what data we had on Japan and some other countries where there have been recent increases in production. We thought this ought to give us some good ideas about the relationship between fertilizer use and crop production. Here, sir, we have a series of graphs showing the correlation between these two variables for a number of countries."

"Yes," replies the Minister, "it's quite evident that fertilizer is important in increasing crop output. Now, how much are we going to need?"

The researcher continues, "According to the latest census, which unfortunately is several years old as you know, only about 40 percent of our farmers are using any fertilizer. This is considerably below the rate in the other countries I mentioned. And in those countries income per farm family has been increasing rapidly, again

1

demonstrating the importance of fertilizer." The Minister nearly interrupts but lets the researcher continue. "Now if we want to double the number of farmers using fertilizer, we might be able to assume that we need twice as much fertilizer as now."

"Yes, I suppose" replies the Minister, "but what about the land area involved and what about the requirements for the different crops?"

The researcher, thumbing back through his report answers, "We don't have any information on area of each crop that is fertilized but cotton and sugarcane consume about 80 percent of the fertilizer used and . . ."

He is interrupted by the Minister who says, "But don't you have any estimates of the quantities required to get the production we need over the next five years? Our problem is that they want to cut back on fertilizer imports to help domestic fertilizer production just at the time we have to try to increase crop production in a big hurry. We need to know how that could affect our program and how much importation we need to ask for." Turning to the head of the planning group he says, "I thought we discussed this pretty thoroughly that day it came up. What happened?"

"Well, that's right, and I knew we were going to have a hard time convincing them how important fertilizer was in our program and I know we talked about that in my office, didn't we?" the head of planning asks the young researcher.

"Yes, sir," replies the young man, "we knew we had to get you some good information on the importance of fertilizer and that's why I have this information on Japan and those other countries."

"Well," replies the Minister, "that's not going to be much help in solving our problem. We have the meeting with them tomorrow. But maybe we could hold off a decision for a couple of days if you think you can get me the information we need by that time. Why don't you call someone at the experiment station? Maybe they can help you. But you better get going. Don't forget how important fertilizer is to us."

There are several important points in this dialogue, which though fictional, represents a real-life situation encountered much too frequently. The most important point is that after waiting right up to the deadline, the Minister, who is the client, did not obtain the information he needed for a meeting of great importance. As a result, the research costs incurred by the Planning Department for this project yielded little of value. The most serious consequence, of course, is the cost associated with not having the relevant information for the meeting. Although some of the reasons for the unfortunate situation are evident in the dialogue, others are more subtle. Regardless of the reasons, we hope that with this book we can contribute to the more effective use of research resources and help prevent the kind of unhappy discussion as that between the Minister and the researcher.

Applied Research

Research is the orderly procedure by which man increases his knowledge and is contrasted to accidental discovery because it follows a series of steps designed precisely for the purpose of developing information.[1] Knowledge gained by research may be used by man to produce a greater abundance of food and fiber, to lighten the burdens of his labor, or in any number of ways to generally improve his well being. Or new knowledge may simply be added to man's store of concepts about the universe to await application at some future point in time. *Research undertaken specifically for the purpose of obtaining information to help resolve a particular problem is applied research.* For a research undertaking to be applied research it is not necessary that the results (the new knowledge) in fact resolve or help resolve the problem which initiated the project (though hopefully they will), but it is necessary that the research have a specific problem orientation. It is this kind of research — that oriented toward resolution of specific problems — toward which this book is directed.

The development of Mexican or dwarf wheat was the result of an applied research process oriented toward the resolution of a specific problem [72]. A fertilizer experiment oriented toward making recommendations to farmers is another example, as is the work of a government planner trying to estimate the likely supply response to fertilizer for a particular commodity under a proposed new program, or the total fertilizer requirements necessary to reach specified production goals, as in our dialogue. Determining acceptability of a newly developed feed concentrate for fattening hogs in tropical areas or the development of a hand seeder for steep terrain in primitive areas also would be classified as applied research. In general, the research referred to in this book is oriented toward providing useful information to decision makers such as farmers and public administrators.

Applied research, such as that just described, is carried out in all parts of the world — it is a much more widespread activity than basic research which is a necessity but one that only the wealthiest countries can afford. Most applied research is conducted under moderate to severe resource limitations which necessitate efficiency

---

[1]For thoughts concerning the meaning of research and the scientific method see [9, 10, 11, 12, 14, 15, 19, 22, 23, 24, 25, 28, 34, 39, 52, 71, 73, 74, 81, 83].

in the research process. *An effective applied research methodology is directed toward the efficient use of available research resources to maximize the probability of achieving meaningful results to help resolve problems.* Disappointment in the results of applied research — a "So what?" response — in most instances can be traced directly to the use of an inadequate and/or ineffective applied research methodology which failed to correctly identify the problem.

The Book

Perhaps the most critical deficiency in methodology is the failure to adequately identify the specific problem toward which the research is to be oriented (as happened to the researcher in the Ministry). This may result when the researcher uncritically accepts the problem as stated by the client or by his spokesman (the "importance of fertilizer" was not the Minister's problem). Another serious deficiency may occur even after properly identifying the problem. This is the failure to formulate hypotheses and objectives correctly oriented toward the resolution of the problem and to use appropriate analytical techniques (what hypotheses did the Ministry researcher use?). The most critical concepts are the interrelationships among problem identification, hypotheses, objectives, analytical techniques, and resource restraints.

The role of theory, though not developed within the main text, is also critical to applied research. Without discounting the value of practice and experience, the greater the command of theory possessed by the researcher, the broader will be his capabilities and the more efficient he will be in planning and conducting the applied research project. This is true because theory envelopes and supports the entire research process.[2] Without a good command of stress theory an engineer cannot properly design nor efficiently build a safe bridge. A plant breeder must understand the theory of genetics before he can hope to efficiently develop a strain resistant to a certain disease. An agricultural economist cannot determine an optimum farm organization without knowledge of production economics theory.

The researcher's foundation in theory provides the orientation for defining a problem that is researchable within the discipline or disciplines involved in the research and with the resources available. Theory also provides the basis for the formulation of hypotheses and in the selection of the analytical techniques to be used. And it

---

[2]For literature concerning the role of theory in research see [20, 35, 40, 53, 62, 63, 70].

should be obvious that the interpretation of the results depends heavily on the theoretical orientation of the researcher.

Although theory permeates the entire research process, in applied research, frequently conducted under sub-optimum conditions, the researcher's practical experience is equally important. Institutional and budgetary restraints, less than ideal field conditions, poorly trained personnel, inadequate background information and other similar factors have a very significant effect on the research process and therefore must be recognized and dealt with accordingly. Practical experience is invaluable in helping the researcher overcome the obstacles which are so often encountered in applied research. In all phases of the material to be presented, the difficulties associated with sub-optimum research conditions under which the individual researcher is apt to be working are considered.

In this book, we have divided the topics into separate chapters and the chapters into two parts derived from the book title; a convention to which we adhere, though not without some reservation. Neither section nor the material in any chapter is independent.

Planning activities are discussed in Part I of the book. In planning the research project, one must always take cognizance of the means available for conducting the research, and during the research process it may be necessary to modify portions of the original plan. Each of the activities is affected by the others and by the research resource restraints under which the researcher is toiling. The kinds and sources of data which will be used and the methods of analysis will be dictated by the hypotheses and objectives, but they, in turn, must be finalized only after having taken into account the effect of resource conditions on availability of data and/or analytical competence.

In Part II, Conducting Applied Research, we discuss experimental and non-experimental data collection, verification and interpretation of data, and presentation of the results to the client.

Because of the deliberate and intense emphasis placed upon problem identification, the book does not discuss in detail each basic element of a research project. One should be cognizant, however, of the interrelated components of a research project and employ them when planning and conducting applied research. These essential elements are: 1) a problem statement accompanied by sufficient information to justify the need for research; 2) hypotheses; 3) objectives; 4) budget; 5) the appropriate theoretical and analytical approach and procedures; 6) data requirements including sources and procedures for obtaining data; 7) a detailed work plan showing jobs to be done and time sequences; and 8) the reports to be issued for each audience.

PART ONE

PLANNING
APPLIED RESEARCH

CHAPTER II

EFFECTS OF RESOURCE AVAILABILITY
ON APPLIED RESEARCH

The relationship of research activities to the availability of research resources is an important difference between applied and basic research. In much of what is commonly considered basic research, a proposal is prepared, and if funding is granted the project is initiated. It normally continues so long as financing is available and then whatever can be concluded is presented as the results of the project. Seldom are the results periodically scrutinized to determine their relationship to the proposal because the urgency associated with solving a pressing problem is missing. Nor are results frequently weighed against the use of resources to estimate the productiveness of the project.

Applied research is more often (though not entirely) carried out under other circumstances. Because the research is oriented toward the resolution of a specific problem, there is usually a **time** restraint fixed by the need to make a decision. Such research is also carried out under varying degrees of **financial** restrictions and usually under rather severe shortages of trained **manpower** and modern **data processing** resources. Another research resource which is seldom abundant under many conditions of applied research is published data, other forms of secondary data or reliable **information** in general. Basic **physical facilities** such as means of transportation or land area for research can also limit the scope of research activities.

In the process of fitting the project to the resource restrictions, three alternatives are possible; 1) the resources may be expanded to fit the project, 2) the project may be narrowed to fit within the restriction, or 3) both may occur within limits. The first alternative is appropriate if the level of precision or degree of confidence desired by the client prohibits a reduction in the scope of the project. In this case, the client must be prepared to provide additional resources where the limitations are critical whether it be in physical facilities, manpower, funds, additional time for completion, or some combination of these provisions.

If resources cannot be expanded for a particular project and the deadline for conclusion is firm, the researcher has four further possibilities open to him. First, he may study fewer variables, ignoring some relationships which affect the analysis, but those which he hopes are less important than the ones included. Second, he may also aggregate variables into groups. In this manner, it is possible to include those relationships which may otherwise have been omitted, but the nature of the relationships, because of the aggregation, becomes less clear. A third alternative is to modify or change the nature of the analysis to be carried out. Less complex analyses can usually be conducted more rapidly and with fewer facilities, but the precision of the results is reduced accordingly. Finally, the researcher may choose to make fewer observations either in the form of fewer replications or a reduced sample.

It is clear that resource availability has an important effect on the nature of the research product derived and the level of precision which can be achieved or the level of confidence which can be placed in the results. Even in very limited resource situations, however, decisions are necessary and researchers are expected to provide useful information. The ultimate decision as to the quantity of resources to be made available for any particular project and the time limit for its completion rests with the client, or with the person responsible for making decisions related to the resolution of the problem toward which the research is oriented. At the same time, the client depends on the researcher to provide him with accurate measures of resource requirements and the scope and precision which he can expect from devoting different amounts of resources to any particular project.

The following sections will describe the principal research resources and include a general discussion of how each resource, when restricted, can affect a project. A single chapter does not adequately cover the range of alternatives open to the researcher and his client, but we hope that it will provide sufficient stimulus so that the researcher, with imagination, will be flexible in adapting his research efforts to any resource situation.[1]

Information Resources

Information is the foundation upon which research is based; hence, one of the major tasks of the researcher is collection of information for use in the research process. Information in general and data specifically are as critical to the problem identification phase of the project as they are to analysis. Their availability profoundly af-

---

[1]See [2] for a discussion concerning allocation of scarce resources to alternative research programs and projects.

fects both the quantity and quality of research which can be produced within a given period of time. There are few limits to the quantity of data that can be accumulated given sufficient time and resources, however, requirements are narrowed and brought into focus by careful research planning.

In general, published information provides the basis for problem formulation while either published data or data generated in the research process (or both) may be the source of information for analysis. Published data, or any information not generated or accumulated under the control of the researcher but utilized in the particular project are considered to be **secondary data**. In contrast, any data generated by the researcher and directly associated with the research project are **primary data**. Another classification of data which has an importance to research is that which differentiates **time series data**, or observations made at specific intervals over a period of time, from **cross-section data** which are taken at one point in time. A third comparison is that between **experimental** and **non-experimental data**. Each of these kinds of data or data sources has different costs associated with availability and analysis, and each has different implications with respect to confidence in conclusions based upon it.

Secondary and Primary Information

When secondary information related to the project is available, its cost to the researcher, both in terms of time and money, is usually less than that required to obtain the same kind of information first hand. However, the usefulness of secondary data as a research resource is not always as great as that of primary information. The researcher must always select the appropriate primary — secondary data mix and the techniques used to combine the two in order to obtain the most useful research product within the limits of his other resources.

For practical purposes, some data are only available from secondary sources. Price series, crop production series and census information are examples. It is simply not appropriate to consider obtaining this kind of information first hand. But it does not follow that available secondary data of this nature are always adequate, representative, or even relevant to the particular project. The researcher must satisfy himself that any particular series really measures something that is relevant to the project or something which can be made relevant through acceptable modifications or manipulation. If care is not taken to verify secondary information in this manner, the researcher may well draw false conclusions from his analysis of the data.

Primary data usually will be more closely related to a particular project than secondary data which are collected for a multitude of purposes or for projects with other objectives. But primary data collection almost always requires more time than is necessary when using secondary data, and may require more of the other resources. Hence, although primary and secondary data are not necessarily substitutes for each other, the researcher should be aware of the availability of secondary information and assess its relevance as an alternative to the collection of primary data.

Time Series and Cross-Section Data[2]

Time series data, as the name implies, are data obtained in a series over a period of time. Examples are price series, production and acreage data, and indices of costs of living or wages, most of which must be accumulated over long periods of time to be useful. Cross-section data are those taken at a fixed point in time (or over a relatively short period of time) and include observations of several different strata or levels of a population. In many cases, time series data are essential to a project, but occasionally, cross-section information can be substituted. Time series data of consumption and income may be used to study the same relationship at a given point in time. Because, as in this case, cross-section data can substitute for a time series, the researcher should not despair if a time series is not available. Nor, by the same token, should he blindly use the time series without considering the advantages of obtaining and using the cross section data. Or in some cases it might be desirable to use a combination of the two.

Experimental and Non-Experimental Data

Experimentation is a means of obtaining data with relatively high precision in measurement of the variables. In many instances this precision is associated with a longer time requirement than that needed for obtaining non-experimental data. In crops a minimum of one season is required and much longer periods may be needed if, for example, effects of weather are to be determined. A year or more may be common for some animal experiments. When, as discussed in Chapter I, the Minister requested that the researcher check with someone at the experiment station to help resolve the fertilizer demand question, he was hoping that secondary experimental data might provide specific crop requirement guides to be used in

---

[2]For a brief discussion of how these two types of data influence demand analysis as one example, see [64].

preparation of a demand estimate. He knew time would not permit the design and analysis of crop experiments.

In some cases, however, experimentation can reduce the time and resources required to resolve a particular problem when compared with non-experiemental data collection. For example, an experiment to determine potential consumer acceptance of a new product before it is marketed can be less time consuming, require fewer resources and involve less financial risk for an industry than consumer response research following the full scale production and marketing of the product.

A possible alternative to experimentation is the collection of non-experimental data through a survey of a number of people who have knowledge or experience with the phenomenon in question. This procedure usually requires less time, but may be less precise and more costly than experimental data collection. An approximation of a fertilizer experiment for example, can be made by surveying a group of farmers, each of whom uses different quantities of fertilizer. Obviously, results will be less precise than experimental results, but at the same time, estimates of variance will be more realistic than from a controlled experiment so that farmers will have a better idea of the range of response to expect from the use of fertilizer.

Another non-experimental source of information in the context of applied research is the simulation of results through the use of "best guesstimates" of the most knowledgable people available to the researcher. In some ways, this is similar to an informal survey. For instance, sufficiently detailed input requirements, yields, and resource restrictions can be generated in this manner for use in preparing budgets, and ultimately a linear program or simulation model for the agricultural sector of a region. Such a model, while less valid than could be possible under more optimal conditions, can be used successfully in project planning where time limits prohibit experimentation and current conditions in the area will not provide the detailed information needed from a survey.[3]

Human Resources

As with most resources, the human element must be considered from the points of view of quantity and quality. Sheer availability is not sufficient for most research undertakings; the training and

---

[3]An example is research conducted in El Salvador [89] by the Food and Resource Economics Department, Institute of Food and Agricultural Sciences, The University of Florida.

capabilities of personnel must be considered when planning the project. Except in rare instances, the time factor in applied research prohibits the training of professional personnel though there may be time to train some non-professionals such as interviewers. A field hand who cannot read or write may be willing to do the physical labor of an experiment, but he cannot be relied upon to maintain records of the results. Nor can professionals with a minimum of training in statistics be expected to carry out complicated statistical analyses. In the first case, other arrangements will have to be made, and in the second, less sophisticated techniques will have to be employed.

A common misuse of research resources exists where elegant data collection techniques are employed but the data are not fully utilized because appropriately trained personnel are not available to make proper or correct analyses. In the short run, simpler experiments and surveys which can be analyzed readily by available personnel are more appropriate. Money saved by not conducting elegant data collection programs may be used to train personnel to conduct and analyze more complicated and sophisticated experiments and surveys in the future.

Physical Resources

Non-technical physical resources include transportation facilities, land, office space, machinery, typewriters, and other items of a similar nature. Like all other research resources, their availability must be considered when planning the project.

Technical physical resources include scientific instruments, calculators, electronic computers, etc. Certain kinds of instruments may be indispensable for particular aspects of a project — some hypotheses may need to be eliminated if the correct instrument is not available or is too costly compared to expected results to justify its use. On the other hand, an electronic computer, while not indispensable, may substitute for other resources such as money or time. The use of a computer, when available, sometimes can shorten the time required to achieve useful results. If a researcher must wait long periods of time, however, for cards to be punched and programs to be de-bugged, he may be better off to undertake appropriate analyses on a desk calculator. When computers and calculators are unavailable a diligent researcher might still provide rough but meaningful recommendations based upon experience and imaginative use of the most basic of physical resources — a pencil and paper.

Financial Resources

This resource to a certain extent can be substituted for all the other resources. At the same time it may not be appropriate, feasible, or even possible to make this substitution because of other considerations. Nevertheless, funds are required for almost all research projects and their availability is an important consideration in research planning.

Time Constraints

Time is usually not thought of as a resource in the same terms as physical facilities, information and human resources, but its effect on the planning and execution of applied research is similar. Considered as a resource, time can interact with other resources in that substitution of one for another can be made. If a decision on a particular problem is critical and time is limited, a greater number of other resources will be required to achieve a given level of confidence than would be necessary if more time could be taken. Hence, the use of more time is an effective substitute for quantities of other resources. But also, consuming more time on one project reduces the amount of that limited resource which is available to help resolve other problems.

In some situations, time can be overwhelmingly limited. This is usually the case when one is involved in so-called "brush fire" research of the type frequently faced by planning groups within the various ministeries of government. In such instances the researcher must always maximize the efficiency with which he uses this resource. Only data readily available can be used and lengthy methods of analysis cannot be considered. Many times "best guesstimates" are the only means available to the researcher under these conditions.

Approaching the other extreme are theses at any of the levels at which they are written in various education systems of the world. Often they are not oriented toward any particular problem so the time factor is not relevant, but among those that are problem oriented, too often time is relegated to a secondary role in resource utilization. As a result, many theses are not written in time to be of any great use — their value reduced by failure to account for the time factor. Those who excuse this fault by emphasizing that theses are only meant to be training tools often deprive the student of an opportunity to perform and benefit from meaningful research.

Summary

Research undertaken specifically for the purpose of obtaining information to help resolve a particular problem is applied research. The purpose of this chapter has been to examine the relationship between the scope of an applied research project and the quantity and quality of research resources which can be devoted to the solution of the problem at hand. The urgency of problem resolution makes time an important resource or constraint which interacts with the other financial, human, physical, and information resources.

The researcher must be aware of the effect that certain resource limitations can have on his research. This cognizance will improve his research effort by increasing the probability that the proposed project will produce useful results. Projects designed in the absence of this consideration can and frequently do run into difficulties such that the productive potential of the resources utilized is not attained. The result is that less effective information is made available for decision making and problem resolution. Careful consideration of resource availability can help prevent situations where the deadline arrives and the researcher, still engrossed in gathering data or making analyses, has little of value to report to his client.

CHAPTER III

ORIENTATION AND FOCUS OF PROJECTS: RESEARCHABLE PROBLEMS, HYPOTHESES, AND OBJECTIVES

Proper management of applied research requires clear definitions of the goals to be achieved through the project. **If one does not know for what he is striving he cannot hope to effectively accomplish the task.** Orientation and focus of the project include the specification of the problem in terms which make it amenable to research, the formulation of hypotheses which are subject to being tested, and the delineation of the specific objectives which the project should accomplish. The interaction of these three phases and their clear exposition serve as a plan or guide for determining the procedures to be followed by the researcher in conducting the research. Adequate orientation and focus of the project will help assure that it can be completed satisfactorily within the resource limitations facing the researcher and will also serve to explain the nature of the undertaking to the client or the administrators for whom the research is being undertaken.

A Conceptual Model

We have found it useful to consider the process of planning the applied research project as being equivalent to a funnel with a series of filters (Figure 1). Such a funnel is used to reduce a large volume of liquid to manageable proportions. The orientation and focus of a research project serves the same purpose — it reduces a large volume of information to manageable proportions. Extraneous information and ideas are eliminated as foreign matter might be filtered in the funnel. Each part of the project statement — problem, hypotheses, objectives — serves to narrow down the proposal, to bring it into sharper focus, and to filter out surplus or unrelated information to make the orientation more precise. One can consider the size of the lower opening of the funnel as being determined by available research resources (the bottleneck). Hence, the proposal as it finally emerges must fit within these resource restrictions.

The top of the funnel will be the general subject matter orientation of the researcher. The research will be within the area of interest of the individual researcher and usually related to his individual talents. The general problematic situation will fall within these limits, but any problematic situation suggested by a client may contain several researchable problems. Hence, the selection of

14

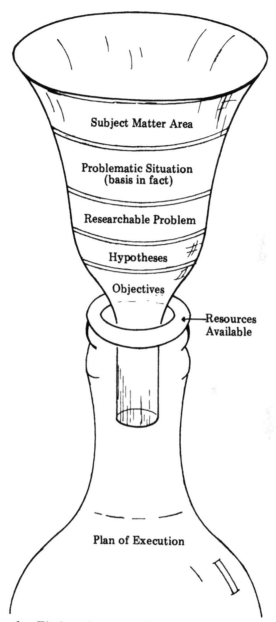

Subject Matter Area

Problematic Situation
(basis in fact)

Researchable Problem

Hypotheses

Objectives

Resources
Available

Plan of Execution

Figure 1.—Fitting the research project to the resources.

a researchable problem based upon the client's needs is equivalent to sharpening the focus on a particular aspect of the more general problematic situation. Hypothesis formulation narrows the problem to tentative relationships which will be tested in the research process. Finally, the objectives specify the limits within which the project will be conducted and describe the useful product which will result.

Obviously, information, ideas and relationships do not flow through the funnel like water but are filtered time and again. The process of moving from the top to the bottom of the research project funnel requires push and pull, paring and adjustment, specification and redefinition until the proper proportions are achieved. The problem, the hypotheses and the objectives may each have to be changed and adjusted many times before a satisfactory product results.

At this point, a word of caution is in order. The result of this funnelling process should be a plan of execution that has a high probability of accomplishing research which will be helpful in resolving the problem toward which the research is to be directed. It is much too common, and seemingly easier, to embark on the next steps of the research process — data collection, analysis, and interpretation — with a poorly specified project statement. The consequence is usually that budget and time restrictions cannot be met and one of the following results:

1) Conclusions must be drawn on the basis of inadequate evidence,
2) More resources and more time must be devoted to the project to allow completion, or
3) The project withers and dies and is relegated to a drawer in the file, never to be heard from again.

No matter what fate the research meets, the return on the investment in the project will be lower than necessary. To avoid these consequences, careful orientation and focus are essential and can well be the most productive time spent on the entire project.

Specification of a Researchable Problem

Problem specification, or elaboration of a problematic situation in such a way that it presents a researchable problem, is a vital step in the process of applied research [42, 84]. Seldom is a client's problem defined for the researcher so that the requirements of the research process are obvious. Even in cases where it may at first appear that it is so defined, it usually is not the case. A common and deceptively simple appearing example is the problem of determining crop production costs. An economist cannot uncritically accept a "cost of

16

production" project without understanding which specific cost components are of interest to the client. The researcher, because of his training, will usually have a greater appreciation for the technical characteristics of the problem than will the client and should therefore consider it part of his responsibility to identify symptoms and diagnose the problem. He should perceive the identification of the problem as a major task.[1] On the other hand, the researcher cannot assume that he automatically understands the problem better than does the client. The researcher must work with the client until they have jointly defined an acceptable and researchable problem.

Problem specification is not a simple process. Hildreth and Castle summarized a discussion concerning problem identification as follows:

> "The start of research is the most important and difficult stage of research. It requires far more than logic; it includes procedures which cannot be neatly categorized and communicated." [56]

But there are several characteristics which a project statement will possess if it defines a researchable problem within the context of applied research. These characteristics of a problem statement are that problems reflect felt needs, problems are non-hypothetical, problems suggest meaningful and testable hypotheses, problems are relevant and manageable, and a researchable problem differs from a problematic situation.

Problems Reflect Felt Needs

A problem exists when there is a need felt by a client. This client may be an individual, a group, or a society. The need must be "felt" in the sense that the originating party believes that change can be realized, and it may arise from social tensions, doubts, conflicts, failure to realize goals, concern for an anticipated occurrence which might be preventable, or the simple lack of knowledge if this knowledge is necessary to contribute directly to the resolution of another felt need. To be appropriate as a researchable problem, the need must be amenable to change as a result of the information supplied by the research process. Hence, all felt needs are not necessarily functional so far as permitting the formulation of a researchable problem statement.

Problems Are Non-hypothetical

A researchable problem statement must be based on factual evidence. The relationships expressed can be neither hypothetical nor subject to doubt or question in the mind of the researcher. The

---

[1]For a discussion of various attitudes toward problem identification in research see [68, 77, 79].

17

researcher must use his judgement as he sifts through available information to determine to his satisfaction what are and what are not acceptable facts and factual relationships. In a probability sense, of course, few facts can be accepted with complete confidence. However, the facts or factual relationships accepted by the client and the researcher in the problem statement must be such that testing of their validity is unnecessary. If a particular proposition cannot be accepted as fact, then if it is relevant to the case, it must be relegated to the status of a hypothesis.

All researchers and their clients will not accept the same information as facts or factual relationships because individual judgements, knowledge and experience affect this choice [5]. In defining the researchable problem related to a shortage of manufactured dairy products, one researcher, because of his experience and general knowledge, may be willing to accept the proposition that there is no shortage of milk processing equipment in a particular area. For him this forms part of the problem statement. Another researcher may feel that the validity of this proposition is not clear, so in the second case, it must become one of the hypotheses (if it is relevant to the problem). Obviously, the nature of the research project will be different in each case.

This is the stage of the research process in which a review of literature is most productive. By means of this review the researcher seeks to utilize the experience of others to help in the orientation of the project. It should be apparent that this kind of orientation can change the focus of the research effort and can have a significant impact on the productivity of the research resources.

Problems Suggest Meaningful, Testable Hypotheses

Because the statement of the problem serves to orient the entire research process, it must suggest testable hypothetical relationships. Hypotheses are formulated as partial explanations of the unknown relationships which create the problem, and those which cannot be tested will be of little assistance in the resolution of the problem. Hypotheses are testable when information about their validity may be collected and analyzed.

The hypotheses must also be developed from the problem statement in a manner which does not result in trivial solutions. Triviality indicates a tautology, an obvious solution, or an infeasible solution.The hypothesis that "per capita consumption of food products is low because there are too many people," derived from a problem statement referring to the existence of hunger in a country is such a hypothesis. Even if the hypothesis is substantiated, it results in an answer which is of little or no use in the alleviation of the immediate felt need.

18

If the problem statement does not suggest testable hypotheses for resolution of the problem under investigation, the researcher has not adequately formulated the problem for research.

Problems are Relevant and Manageable

Agricultural scientists tend to work at extremes. "They tend to work either on problems where the outcome is highly predictable but which has little impact on problems or on problems so large as to be unmanageable" [56, p. 38].

This comment was made with reference specifically to agricultural economists but it applies to other disciplines as well. An agronomist designing an experiment to determine yield response on a certain soil, even though information about similar soils is readily available, is an example. The researcher knows with a high degree of certainty whether or not a response will be observed.

Overambition, lack of adequate forethought, and inexperience are the principal causes of unmanageable projects. An over-ambitious researcher may be motivated by a desire to study all the problems in a given sector so that he can answer any question that arises. Or an unmanageable research project can result from suggestions by clients or administrators unfamiliar with the discipline of the research. In this case the researcher should not accept the project without first more precisely defining the problem and reducing the proposed project to manageable proportions within the time requirement and within the human, financial and other resources available for the project.

An unmanageable research project yields few benefits to anyone. The most likely result of such a project is superficial treatment of part of the problem and the neglect of other parts. Little detail will be achieved, and much that is presented may be found to be inaccurate or insufficient. The contribution of this type of study will almost always be less than that of a more specific study which examines fewer phenomena, but does so in more detail.

Researchable Problems vs. Problematic Situations

Confusion is likely to exist relative to the differences between a problematic situation and a researchable problem. In our context, there is a real and functional difference between the two. First, a problematic situation is a phenomenon which exists; a researchable problem must be identified and defined. A problematic situation represents a generalized situation but a researchable problem, expressed in the above terms, must be specific. "An increasing rate of crime in the cities," assuming the statement is true, is an example of a problematic situation — it exists and it pertains to felt needs. But in no way can the statement be construed as a statement of a researchable problem. The expression of a problematic situation can serve as general orientation to the broad area of interest of a research project, but it is not sufficient to serve as a guide to a research project.

A problematic situation can be the source of a variety of researchable problems. Different clients, different research administrators and different researchers are likely to arrive at a variety of researchable problems from the same problematic situation. That is to say that there is no single right, or correct, researchable problem — there is only a correct framework within which the researchable problem must be expressed to be useful to the researcher. The selection of **the** researchable problem which will actually be undertaken will depend on the situation within which the research is being conducted.

Examples of Problem Statements

In evaluating statements which are considered by their author to delineate researchable problems, it is difficult, if not impossible, to determine when deficiencies are due to lack of care in problem formulation on the part of the researcher or are due to an unclear concept of the nature of the problem. An example of such a statement is the following:[2]

1) "Rural unemployment created by an increase in the use of agricultural machinery."

There is no doubt that the author was focusing on a problematic situation and his interest had to do with unemployment. But it is not clear what specific problem existed in his mind. The statement does not provide the basis for a research project though it might spark a lively debate. As it stands, it does not meet the basic requisites for the specification of a researchable problem. Other examples of incomplete or incorrectly formulated problem statements are the following:

2) "Few cattlemen vaccinate against hoof and mouth disease."

3) "Exportation of fresh meat from Colombia."

Let us analyze the above three statements from the point of view of the requisites necessary for a correctly specified problem statement. Do they concern a felt need? Probably all do, but what exactly, is that felt need? The felt needs suggested by those statements

---

[2]The examples used in the remainder of this chapter are taken from class papers and research proposals submitted by graduate students in the Department of Agricultural Economics, Instituto Colombiano Agropocuario, Bogota, Colombia.

are numerous. Are the rural unemployed a problem from the standpoint of crime or poverty in the cities, or is the author perhaps considering those unemployed as a source of inexpensive labor for rural industry? Is the author (or client) of the second statement the president of a drug company or is he the head of a meat export company? Is the third statement a problem from the point of view of Colombia, which may be looking for increased foreign exchange, or, perhaps of Argentina which may be considering Colombian competition in the world market? Each of these points of view suggests a different problem, and hence, a different research project. It would be folly to initiate a research project on any of the above topics without more complete specification of the client's felt need.

Are the statements hypothetical or subject to question in the mind of the author? This is the most difficult test for a third person to make because he is not aware of the precise knowledge held by the researcher or client. For this reason, proper background information in the research project statement is critical. Not everyone will automatically be convinced that the increased use of agricultural machinery leads to rural unemployment. However, if such a relationship is acceptable to both the client and the researcher, and is considered firm and non-hypothetical, it can be acceptable in the problem statement as a non-hypothetical relationship. The second and third statements do not express causal relationships and hence are not subject to this criterion. They are simply incomplete.

Do the statements suggest testable hypotheses? The first statement could be formulated as a testable hypothesis, but if it is to be accepted as a factual relationship for a problem statement, then it cannot be a hypothesis for the purposes of the research. Although testable hypotheses associated with the other two statements could be contrived, any firm relationship between them and the problem in the mind of the reader would be purely coincidental.

Another problem statement will be analyzed to determine if, or how, it might be improved.

4) "Deficient milk production reduces domestic consumption and makes imports necessary in this agricultural subsector."

On the surface this appears to be a more complete problem statement than any of the preceding . Certainly it represents a felt need; in fact more than one. Is the problem to which the author is directing himself a production problem, one of poor nutrition because of inadequate quantities of milk, or is he concerned with the expenditure of scarce foreign exchange? From the present statement, it is impossible to be sure.

21

The relationships expressed probably meet the non-hypothetical requisite, though not necessarily. The first relationship is really a tautology if "deficient" is defined in terms of domestic consumption. The second relationship could be hypothetical unless government policy, for example, has provided for the imports of milk to make up deficits in domestic consumption.

The statement does not legitimately suggest any hypotheses except the trivial ones that increased production would increase consumption and/or reduce the necessity for imports. And these are not really testable in the usual research framework. Hence, even though the problem would be amenable to one or more meaningful solutions, it is probably not researchable within the usual limits of time and funds. Several years would be required to determine experimentally if increased domestic production would, in fact, result in increased consumption.

In order to improve this statement, it is first necessary to focus more precisely on the orientation of the author with respect to his felt need. It turns out that the author of the statement was concerned with problems of production, principally with high costs, low productivity, and the small profit margins of the producers of milk. His research interest then focuses on the dairy herd and the dairy farm, and possibly on the farm-to-market process. Although it will never be possible to determine the exact nature of a proposed research project from the problem statement alone, a more precise statement is necessary before one can proceed to formulate hypotheses and specify objectives. A better formulation of the above statement resulted from more reflection by the author on his orientation:

5) "The low level of technology on dairy farms contributes to high costs of production, low average productivity, and a deficient marketing system for milk. These factors cause low profits to the producer, price fluctuations for the consumer, and a deterioration in the balance of payments because of the necessity to import milk."

In this form, the statement implies, in accordance with the requisites, that the author accepts as fact that a low level of technology is a causal factor in low productivity, high costs of production and a deficient marketing system for the product. Furthermore, he accepts that these factors cause low profits, price fluctuations and the necessity of importing milk. By accepting these relationships as fact, they cannot appear as hypotheses to be tested by the research process. If the researcher or his client cannot accept any of these relationships as factual, then the problem statement should be modified and the doubtful statements submitted as hypotheses to be tested (assuming they can be tested within the limits of the available research resources.)

Formulation of the Hypotheses

It is clear that the logical sequence of events in the process of applied scientific inquiry begins with the observation of phenomena in the empirical world in common sense terms. Determination and classification of these events into problematic situations and specific researchable problems set the stage for postulating various potential means of problem resolution. This part of the process involves the formulation of hypotheses which formalize the premises to be tested by the research effort.

Hypotheses are tentative propositions that must relate to the problem so as to be helpful in providing means of resolution. Any suggested or indicated means of resolving a problem must be formulated so as to be testable and the relationship to the problem must be evident. To be logically complete and functional a hypothesis must involve a relationship. Such a formulation constitutes, implicitly if not explicitly, a hypothesis in the form of if-then propositions. The "if" clause describes the relationship between the postulated condition and the proposed result. For example, "If Colombia can increase its beef production by 15 percent and reduce the cost of production by 10 percent, then it can successfully export beef and meet domestic requirements." The first clause "If Colombia can increase its beef production by 15 percent and reduce the cost of production by 10 percent" sets the conditions that must be met and the remaining clause relates the proposed results.

Hypotheses are derived from the observations and relationships accepted as, or assumed to be, fact in the problem statement. They provide the guidelines for the type of data and techniques necessary for analysis. This implies that hypotheses are formulated before the data collection activity of the research project has started. In this sense hypotheses indicate the direction for data collection; hypotheses that are formulated to explain observations after they are collected may not be useful for problem resolution. The hypotheses, then, are a necessary link between the problem and the data collection and analytical stages of the research.

The basis for correct formulation of hypotheses is the knowledge of the researcher, this knowledge being founded primarily in theory.[3] The broader the experience of the researcher in relating theory to applied problems, the more efficient he will be in formulating appropriate hypotheses. Whereas the researcher and the

---

[3] For this argument and others relating theory to the research process see [35, 45, 53, 63, 73].

client, jointly, must share the responsibility for problem specification, it is primarily the responsibility of the researcher, as the expert in his field, to formulate the hypotheses.

Characteristics of Hypotheses

Hypotheses appropriate to applied research have the following characteristics:

1) They must be formed as **if-then** relationships and stated in such a manner that their implications and relationships to the problem can be shown logically. The explicit use of the words **if-then** is not necessarily required; the relationship, however, is critical and often an explicit **if-then** statement will assure an accurate relationship.

2) They should be stated as simply as possible both in terms of theoretical complexities and implications and in terms of number of variables.

3) They must be capable of verification or rejection within the limits of the research resources.

4) They must be stated in a manner which provides direction for the research. The hypotheses, when well formulated, will suggest the appropriate data and analytical techniques for testing that should be employed in the research process. Thus, a set of hypotheses can be thought of as a plan of action.

5) Taken together, they must be adequate and efficient in suggesting one or more meaningful solutions to the problem. They must provide for an acceptable level of confidence in the results, but at the same time economize in the use of scarce research resources.

Some Examples of Hypotheses

To continue with the proposed dairy study, the following are the hypotheses which were submitted in the second draft of the proposal:

1) "Increasing the level of technology and of physical production and the economy per unit of exploitation, in association with a reduction in the costs of production, would result in stability between supply and demand."

2) "Providing more financial resources for increasing production and restructuring the market channels would allow simpler price regulation."

3) "Establishing milk regulations would provide optimum quality and a price warranted by that quality."

Although all the hypotheses are generally related to the problems as stated, it is quite obvious that they encompass a much broader front than we were imagining from the reformulation of the problem statement. This revelation leads one immediately to the suspicion that the hypotheses violate the second requisite of hypotheses — simplicity. The first hypothesis encompasses a fair proportion of all economic theory and certainly does not clearly specify how one moves from supply to demand with ease. Any reasonably low-cost means of testing this hypothesis is hard to imagine. The second and third hypotheses are apparently related to the problem through the concern with price fluctuations and, perhaps with the low profits of the producer. However, the relationship to the problem is only tenuous at best, and actually introduces new concepts that were not in the problem statement. Rather than helping to clarify the research proposal, as they should, these hypotheses tend to add confusion. Clearly, they do not provide direction or guidance for the research which is one of the primary functions of hypotheses.

Because the author of the hypotheses in the example did not progress beyond this point, we must now begin to act as if we were the researcher and formulate the hypotheses according to our understanding of the problem. In doing so, it may be necessary to better identify the problem itself (normally done in consultation with the client) so that the proposal can be improved. Since this occurs many times in the formulation of a good research proposal, we will not be wasting effort to do it at this point. Few people are capable of writing an acceptable research proposal on the first attempt. Several modifications usually are necessary as the orientation and focus become clearer. The final version of the problem statement in the previous section read as follows:

> "The low level of technology on dairy farms contributes to high costs of production, low average productivity, and a deficient marketing system for milk. These factors cause low profits to the producer, price fluctuations for the consumer, and a deterioration in the balance of payments because of the necessity to import milk."

Let us assume that we were correct in deciding that the orientation of the proposal was more toward production than toward marketing. Let us also accept the relationships expressed in the problem statement although many of us may not be comfortable with some of them. Then, presumably, the focus of the proposed research project is the low level of technology on dairy farms, this being the cause of the unfavorable relationships expressed in the problem statement.

Because the hypotheses must be testable within the resource limitations, care must be exercised in comparing resource requirements and limitations before continuing with the formulation of the proposal. Assume that the client in this case is the Planning Department of the Ministry of Agriculture. In preparing their plans for the following year, they are considering the long range necessity of importing milk. They have requested a series of studies to help them clarify the situation so they can make firmer predictions and to aid them in establishing domestic policies. They consider that the country is capable of producing more milk but are not sure why it is not doing so. One factor that is evident is that there is a low level of technology used in the dairy industry. In this particular study their desire is to determine why technology has not improved in recent years, because general knowledge far exceeds implementation. Professionals will be provided to work full time on the project, and the Ministry wants a preliminary report in four months and a final report in six. Generally, calculator and computer facilities are available and it will be possible to hire the services of some interviewers for a short period of time if necessary.

The first hypothesis might well treat the profitability of the new technology because, if it is not profitable, most assuredly the producers will not adopt it. For a beginning, let us use:

1) "There exist in the country improved methods of dairy production which, if used by producers, would increase their profits."

Stated in this fashion the hypothesis does indicate the possible direction which the research might take. In verifying or rejecting it, one method might be a simple review of literature to determine if we can satisfy ourselves as to the existence of profitable new technologies. Perhaps the technologies which are available have never been subjected to tests of profitability, which would indicate the need for some partial budgeting or possibly linear programming for a series of typical farm resource situations. Of course it might also be true that the client is satisfied that there are profitable new technologies and therefore does not desire or require verification of this hypothesis. Assuming that we will need to include this hypothesis in the project, further clarification will be made in the statement of the objectives and the procedures.

The hypothesis can be stated in an **if-then** form, "If farmers adopt presently known technology then their profits will increase," and the relationship to the problem statement is clear. Hence, it satisfies the first requisite of hypotheses.

Depending on the other hypotheses which are derived, all of the previously discussed means of verification or rejection of the hypothesis could be accomplished under the limitations of the budget which we have described.

The theoretical implications of the hypothesis are about as simple as it is possible to make them and still retain meaningful relationships to the problem. To be profitable, the new technologies must fit within the resource restrictions of the farmer and must be profitable given present or projected price situations. Notice that it is not adequate, even though it is simpler, to hypothesize only that the new technologies could increase farmers' production.

The last criterion of hypotheses applies to all of them taken together, so it cannot be applied individually except in the sense that it would not be adequate to consider only an increase in production without considering, at the same time, the profitability of the practice.

Assuming that in the course of the research we will be able to accept the first hypothesis, we must then consider additional hypotheses because the first, alone, is not adequate to answer the question put to us by the client. The second hypothesis might be that,

2) "Farmers are not aware of the new technologies."

As stated this is a known fact, but the **if-then** implication is that if farmers are not aware of profitable alternatives then they will not (or cannot) adopt them. The direction of the research in this case is also indicated. A sample survey of farmers should be able to provide the evidence necessary to accept or reject the hypothesis.

Another hypothesis could treat credit, or the related financial situation of the farmers.

3) "If special credit sources are not made available, farmers will not be able to adopt the new technologies."

This can be considered a broad hypothesis with two sub-hypothesis which will be tested,

a) "Farmers are unable to adopt new technologies because internal financial resources are limited."

and

b) "Farmers are unable to obtain credit, which limits their ability to finance changes in their production techniques."

An additional hypothesis could treat the problem of price instability from the point of view of the producer.

4) "A milk price stabilization program could induce farmers to adopt improved methods of production."

This hypothesis is not quite as straightforward as the others. To be able to verify or reject it empirically, a great deal of time and money would be involved — certainly more of both than we have available within our research budget. However, attitudes of farmers toward a price program could be ascertained and certain conclusions drawn regarding the **probable** outcome of such a program. Again, if this hypothesis is included, the precise nature of the research process will need to be specified in the objectives and the procedures.

27

Other possible hypotheses could be suggested, but those proposed above are adequate for present purposes. Probably, all can be accomplished within the research budget, but it will be necessary to clarify them further (and perhaps modify them) as we develop our statement of objectives and procedures.

In summary, our hypotheses are the following:

1) "There exist in the country improved methods of dairy production which, if used by producers, would increase their profits.

2) Farmers have not adopted the new methods because they are unaware of their existence.

3) Special credit sources are necessary if dairy farmers are to adopt improved methods of milk production.

   a) Dairy farmers are unable to adopt new technologies due to internal financial resource limitations.

   b) Dairy farmers are unable to obtain credit, which limits their ability to finance changes in production methods.

4) A milk price stabilization program could induce farmers to adopt improved methods of production."

Are these hypotheses, taken together, adequate and efficient in suggesting a means to one or more meaningful solutions to the problem? If they cover the range of possibilities open to the government (extension programs, credit programs, and/or price programs, for example) they should be adequate in suggesting guidelines to one or more meaningful solutions. They are efficient if we cannot contrive other hypotheses which could provide solutions to the problem with the use of fewer of our research resources, do so in less time, or result in more precise information within the research budget.

Delineation of the Objectives

Objectives usually are expressed in lay terminology and are directed as much to the client as to the researcher. The primary objective of applied research will be either 1) to suggest or recommend to the client practical means of problem resolution, or 2) to provide information to clarify an unknown situation. Generally, the objectives taken as a group will 1) define the limits of the research project for the researcher, 2) clarify the means of conducting the research, 3) identify the client or clients, and 4) describe the expected product of the research for the client.

The objectives link the theoretical relationships presented in the hypotheses to the analytical and methodological orientation necessary for conducting the research. An objective specifies what the researcher intends to do or find in the project and suggests one or more research procedures to be used. Later in the research proposal the specific procedures must be defined. These procedures of course, must be appropriate, not only to the stated objectives but also to the resource availability as discussed in Chapter II. It is necessary, therefore, that the objectives be sufficiently broad to satisfy the needs of the client but also sufficiently specific to conform to budgetary restrictions.

An erroneous idea of the nature of the objectives of a research project should be clarified. Research objectives are neither political objectives nor are they objectives of an action program of the government. Objectives of a research project suggest what information will be obtained for the client to help resolve the problem which initiated the research. This information, in turn, can be used by the client to establish an action program. For example, "to redistribute land in Mexico" suggests a government action program. This cannot be an objective of a research project. But, "to study the past and present agrarian reform programs in Mexico and their effects on land redistribution" is a legitimate research objective related to a felt need concerning land use and productivity. In the terms of our dairy problem, "to establish a credit program for dairy farmers" is not a legitimate research objective, though it may be a feasible solution in the mind of the researcher and the client. "To determine if dairy farmers are able to obtain credit for improving methods of production," or "to recommend methods of resolving credit deficiencies if identified," on the other hand, are acceptable research objectives.

Given these conditions, the primary objective of the dairy study which we are proposing could be, "to determine the obstacles to the adoption and optimal use of new technology by dairy farmers." This objective precisely describes the purposes of the project. Accomplishing this objective will provide information to clarify an unknown situation for the client, and, if we are successful, should provide him with the information he needs to make policy decisions and predictions concerning future milk supplies.

Secondary objectives can include the following, listed in order of their relationship to the hypotheses:

1) Determine if presently known modern technology is profitable to the dairy farmers given their present resource situation and market outlet.
2) Determine if the Extension Service is effectively providing necessary information to farmers concerning possible alternatives for production.
3) Determine if the required changes to adopt new technologies on farms are outside the financial means of the farmers.

4) Ascertain whether present credit sources are adequate in providing for the farmers' needs related to new practices.

5) Obtain farmers' opinions about a price stabilization program and their possible reaction to it with respect to changes in use of technology and concurrent production practices.

Notice that each of these objectives is directly related to a hypothesis and help to clarify the direction of the research. If the client does not understand the terminology used in the hypotheses, he should be able to understand the nature of the research from reading the objectives. Also, the objectives clarify the means of conducting the research. It is clear that interviews with dairy farmers will be necessary and, in general, what the content of these interviews will be. Specific questions must be written from the guidelines given in the objectives. These questions either singly or as groups will provide the information to test the hypotheses.

Of special interest is the fifth objective. The related hypothesis (a milk price stabilization program could induce farmers to adopt improved methods of production) states a positive relationship without giving any indication of the nature of the test which the researcher has in mind. The client in such a case could well formulate an erroneous impression of the nature of the results which the researcher proposes to present him. The fifth objective specifically states that the researcher expects no more than to obtain the farmers' **opinions** and **possible** reactions. This is clearly different from empirical evidence which the client might otherwise expect if he only had the hypothesis for information.

We have been talking about a special or explicit client but other clientele could also benefit from the research in question. If the researcher is hired specifically by a client, and the client considers the research private, then, of course, he is the only one who will directly benefit from the work. However, as is more often the case, this type of research is undertaken by a public or semi-public entity so the research becomes public property and the identification of a broader audience is useful. In the context of the present problem, the full clientele could be identified in a sixth objective:

6) Provide information to farmers about the profitability of new methods of production, to bankers and other credit institutions of possible sources of new business, and to government planners to aid them in making decisions concerning the future of the dairy industry.

Summary

Three of the most important and critical aspects of the planning of a research proposal — problems, hypotheses, and objectives — have been presented and discussed. These three parts are not independent from each other, nor are they independent from other portions of the proposal and aspects of conducting the research to be discussed in following chapters. They serve as a framework for developing the data collection and analytical procedures, the budget including time sequences and the publication plans for the research results.

Time spent in careful development of the problem statement, the hypotheses, and the objectives is the key to efficient research and can well be the most productive use of time by the researcher. Even in cases where the researcher may have only one month, one week, or perhaps just one day to provide an answer, the time spent in this phase of the research is critical to the success of the undertaking. On many occasions when a person is given a rush task, the tendency is to "come up with something." Little time is devoted to analyzing the situation to determine precisely what the client wants, what is really needed, and what resources are available to accomplish the necessary task. Too often the results have no value because the "something" which the researcher "comes up with" is not really related to the problem of the client.

Problems appropriately specified for applied research have the following characteristics:

1) They are based on felt needs of individuals, groups, and societies;
2) The causal relationships expressed in a problem statement are not hypothetical and are relevant to the problem;
3) Problem statements must suggest testable hypothetical relationships that, when analyzed, yield relevant and nontrivial results;
4) The problem and the research to resolve the problem must be relevant and manageable within resource restrictions.

Researchable problems can be distinguished from problematic situations in that numerous researchable problems can be formulated from a problematic situation.

The hypotheses serve as guides to executing the research. Hypotheses must:

1) Be stated to provide direction for the research;
2) Be formulated as causal relationships with **if-then** implications;
3) Be capable of tests within the limits of the research resources;
4) Be stated as simply as possible; and

5) As a group be adequate and efficient in suggesting means to one or more meaningful solutions to the problem.

Objectives in general describe what is expected to be achieved by the project. Specifically, objectives:
1) Define the limits of the research project;
2) Suggest or clarify the means of conducting research;
3) Describe the nature of the potential research product to the client; and
4) Identify the client or clients.

The following problem statement, hypotheses and objectives are those of the example discussed in the chapter.[4]

Problem:

The low level of technology on dairy farms contributes to high costs of production, low average productivity, and a deficient marketing system for milk. These factors cause low profits to the producer, price fluctuations for the consumer, and a deterioration in the balance of payments because of the necessity to import milk.

Hypotheses:

1) There exist in the country improved methods of dairy production which, if used by the producers, would increase their profits.
2) Farmers have not adopted new methods because they are unaware of their existence.
3) Special credit sources are necessary if farmers are to adopt improved methods of production.
   a) Farmers are unable to adopt new technologies due to internal financial resource limitations.
   b) Farmers are unable to obtain credit, which limits their ability to finance changes in production methods.
4) A milk price stabilization program could induce farmers to adopt improved methods of production.

Objectives:
1) To determine the obstacles to the adoption of new technology by dairy farmers.
2) Determine if presently known modern technology is profitable to the dairy farmers given their present resource situation and market outlook.

---

[4]See the appendix for an example of a complete project statement.

3) Determine if the Extension Service is effectively providing necessary information to farmers concerning possible alternatives for production.
4) Determine if the required changes to adopt new technologies on farms are outside the financial means of the farmers.
5) Ascertain whether present credit sources are adequate in providing for the farmers' needs related to new practices.
6) Obtain farmers' opinions about a price stabilization program and their possible reaction to it with respect to changes in use of technology and concurrent production.

The chapters to follow will continue with discussions of data sources and collection, and analysis and presentation of research results, all of which depend upon well specified problem statements, hypotheses and objectives.

PART TWO

CONDUCTING
APPLIED RESEARCH

CHAPTER IV

EXPERIMENTAL DATA COLLECTION

The execution of research involves the collection of data which pertain to the project, the utilization of the data to test the hypotheses (usually called analysis of the data), reaching conclusions which are useful in the resolution of the problem which initiated the research, and making appropriate recommendations to the client. If the project has been properly planned, the type of data which are needed have been determined prior to conducting the research. The researcher will know if only secondary information will be used or if primary data collection will be necessary. The planning phase of the project will also determine if experimentation is necessary or if the source of data will be non-experimental. If experimentation is to be used, a great deal of care must be exercised in experimental design.

Experimentation has been the principle basis for obtaining scientific information and will continue to remain of paramount importance. The use of non-experimental data, however, is becoming more prominent, due to improvements in measurement techniques and development of more adequate data series. The advantage of experimental data over non-experimental data is basically the degree of control which the researcher is able to exert over the variables included in the study. In most experiments, with the use of the proper design the researcher can select the factors which will vary, the levels at which they will be included in the experiment, and the pattern in which they will be used. By using appropriate equipment he can also usually obtain quite accurate measures of the input variables as well as of the results of the experiment. With non-experimental data, the levels and combinations of variables are predetermined by nature or society, so the researcher must measure and use them as they exist. Because these variables are very difficult to identify and measure, non-experimental data are usually more subject to inaccuracies than are experimental data.

Experimental Design

Experimental design, the form in which the experiment is to be set up, is a critical factor in the generation of experimental data.[1] There are any number of designs which can serve a particular need, but there will usually be one that is better suited to the conditions under which one experiment is going to be conducted and which is the most **efficient** in the use of the research resources. Selection of the most efficient experimental design will provide more information directly related to problem resolution for the given set of research resources than any other alternative design.

Selection of the number of treatments, the number of controlled and measured factors, the levels at which the variables are to be included, and the number of replications to be used in the experiment is not a simple process. The final choice can be complicated in applied research because the conditions under which the researcher is working are often poor in relation to what he would like for improved precision. Good research planning is extremely helpful to the researcher in choosing an experimental design. This is so because the design must be related to the problem and the methods of analysis to be used and these should have been carefully considered in the planning phase of the project.

In the choice of a design, then, the researcher will be able to anticipate the type of information which will be forthcoming from any particular design. He will be able to predetermine the applicability of the design to the problem and to the methods of analysis which are appropriate. The researcher will also ascertain whether or not the experiment can be conducted within his resource limitations.

Relationship to the Problem

Although it is possible to achieve measurement accuracy with experimentation, it does not necessarily follow that experimental data as a source of information for a research project will assure research precision. If the design of the experiment is not properly related to the problem orientation of the project it will not be possible to achieve precision in determining relationships useful to problem resolution. An example may best serve to illustrate the point.

A common type of design in fertilizer experimentation is the following, with each treatment being replicated a number of times (frequently four):

---

[1]Selected references on experimental design include [3, 4, 7, 26].

Design 1

Treatment Number	Variables		
	N	P	K
1	0	0	0
2	0	0	1
3	1	0	1
4	2	0	1
5	0	1	1
6	1	1	1
7	2	1	1
8	1	1	0

In the project proposal an objective might read something like, "to determine the effect of nitrogen (N), phosphorus (P) and potassium (K) on the production of pangola grass in the Cauca Valley." As we now know, of course, part of the difficulty here is that we do not know what the identified research problem is. But that aside, the objective as stated is vague and is not an adequate guide for designing the experiment. Toward what end is the experiment oriented? Does the researcher want to know simply, "is there a response to N, P, and K?" — or is he more interested in the magnitude of the response, or even the kind or form of response over a range of N, P, and K in various combinations?

Each of these questions may well require the use of a different experimental design, number of treatments or number of replications in order to achieve precision and efficiency in arriving at an answer. In some cases, two or more questions can be answered efficiently from one experiment, but in others, the attempt to answer too many questions from a single design may render the experiment useless for answering any question.

Without discussing the theoretical logic for the statements, the following can be said about the relationship of Design I to some of the questions which could be asked. When the researcher is interested in knowing only if there is a response to each of three nutrients, it is not necessary to include all the treatments and levels of N in the design. A simpler design such as the following could be used:

Design II

Treatment Number	Variables		
	N	P	K
1	0	0	0
2	1	0	0
3	0	1	0
4	0	0	1

Here, instead of eight treatments there are only four. Using four replications of each treatment, only 16 separate plots are required rather than 32 as would be necessary in the first design if four replications were also used. The choice of the level of each nutrient (1 = 100 kg/ha or 1 = 200 kg/ha, etc.) is important because this design provides information only for the one level chosen.

If the researcher wanted to know the magnitude of the response for two different positive levels of N, it would be necessary to include the three different treatments (0, 1, 2) of this nutrient. Apparently in Design I it was desired to know something about the magnitude of reponse to N for two different levels of P, because the N treatments are repeated for 0 and for 1 P. This question can be answered with the use of treatments 2 through 7 of Design I.

The last treatment of Design I can only answer two questions, 1) "Is there an effect from K when N = 1 and P = 1?" This requires treatments 6 and 8. No other information will be available on the effect of K. The last treatment along with the control (treatment 1) can show the response of N and P together when No K is applied.

Because we are interested in applied research, let's suppose that the ultimate use of the data from the proposed experiment will be to make fertilizer recommendations to farmers. To what extent does Design I permit us to make the appropriate analyses? One alternative is to compare the cost and return of each treatment with the control (0-0-0) and determine the net return from each. Assuming that there were significant differences between treatments, we could then recommend the treatment with the highest net return.

Another alternative would be to determine the functional relationship between N and production (production function for N) for each of the two levels of P. Assuming the curves thus generated were of proper form (satisfying the conditions of optimality), we could determine the economically optimum quantity of N for each level of P, make a cost-return estimate for each, and choose the one with the higher net return as that which we would recommend.

The second procedure would probably result in a somewhat different recommendation than the first method. There is obviously less precision with respect to N in the first method than in the second but there may be even less precision with repect to P because we have only two treatments to choose from. Apparently, the design is not well suited to answer the question the farmer might ask, "What is the best combination of N, P, and K to use?" even though it does appear to be at least reasonably suitable for answering some other questions.

To answer the farmer's question may require a more complicated design and could require more resources than are available, but much more precision is possible if there are sufficient resources. As an example, a complete factorial with 3 or 4 levels of each nutrient would require 27 or 64 plots respectively for each replication, but could provide rather precise answers for the farmer (the use of the 4^3 factorial, of course, would provide more precision than would the 3^3 factorial but it also requires more resources). Another, more efficient design is the rotatable central composite which is a modified factorial that requires only 15 plots for each replication when three nutrients are considered. For this last design, fewer replications are necessary so that with only about 40 plots a complete experiment can be conducted,[2] providing a wide range of information.

In summary, the problem toward which the research is directed has a strong bearing on the type of experimental design which should be chosen if the researcher is going to use experimentation in the execution of the project. Should the magnitude of the problem require a very detailed experimental design, the researcher should be encouraged to consult with a competent statistician when one is available. Also, if the researcher is going to use secondary experimental data, the design which was used in the experiment should be considered when choosing between various sources. More will be said on this aspect in a later section of this chapter.

Relationship to Resources

The ultimate size and complexity of an experiment depends on many factors including number of independent factors to be controlled or measured, analytical techniques to be used, statistical precision required, quantity and quality of prior information available, the objectives of the research project, the time within which results must be obtained, and the resource constraints encompassing the research project.

If it is necessary to determine the best combination of N, P, and K, a rather complex design with a relatively large number of experimental observations (plots) is required. But if it is simply not possible to conduct such a complex experiment, the research project must be modified. For example, if some information is available which indicates little or no response to K, then the number of independent variables could be reduced to two (N and P) and K could be excluded or held constant at some predetermined level in the experiment. The best combination of N and P can then be determined with a rotatable central composite design requiring only about 26 experimental units rather than the 40 units required for three independent factors (N, P, and K).

The design can be simplified further by considering only one independent factor. This factor could be the one considered most im-

---

[2]All treatments are not replicated the same number of times [55].

portant such as perhaps N or it could be some fixed combination of several factors such as a complete fertilizer containing N, P, and K with an analysis of 10-30-10, for example.

An experiment requiring only about 15 experimental units will yield as much statistical precision for one factor as 26 for two factors. But it is important to realize that the total amount of information is reduced accordingly as the number of independent factors and resource requirements are reduced. With three factors and 40 plots, we should be able to tell the farmer, who is on similar soil as that used for the experiment, how much N, how much P, and how much K he should use to achieve the greatest net returns from fertilizer use. With two independent or variable factors, we have the option of telling the farmer the best combination of only two nutrients and can do that only for some predetermined quantity of the third nutrient. With only one variable factor it is not possible to recommend best combinations but only the best quantity of that single factor. We may be able to recommend 300 kilos per hectare of 10-30-10 as the best quantity of that complete fertilizer, but we would not be at all confident that the resulting 30 kilos of N, 90 of P_2O_5, and 30 of K_2O would be the best combination of the three nutrients to use.

The effect on resource requirements of the analytical technique to be used is more complex, but an example will serve to illustrate the point. Suppose that it is desired to determine if there are significant differences in the response of three different levels of a new feed additive on fattening steers. A probable design would have three treatments with the additive (one for each level to be tested) and a control. This results in four treatments per replication. Under most conditions three to four replications have been found to be required to obtain reliable statistical estimates and provide enough degrees of freedom for this type experiment and the analysis of variance which would be used. A total of 12 to 16 experimental units (pens of cattle) would be required to execute the research.

An alternative to analysis of variance is regression analysis. If this technique is envisioned for the feed additive experiment the same four treatments could be used and if the interval between treatments was not uniform, they could be adjusted to make them equal (a desirable though not necessary attribute of a design for regression analysis). In analysis of variance, the observations from only two treatments are compared at any one time. In regression analysis the observations from all the treatments are considered simultaneously. For this reason roughly the same amount of statistical precision can be otained with two replications of the four treatments for regression analysis as with four replications of the four treatments for analysis of variance. Hence, regression analysis may require fewer experimental resources than analysis of variance. Similar relationships hold for other types of analysis which may be required in the project and should be considered when choosing the design to be used.

The statistical precision required in experimental results is associated closely with the nature of the problem for which the research is being undertaken. Lower levels of confidence are acceptable in research dealing with animal health than with human health, for example, and lower confidence levels may be satisfactory in some social research than in some biological research. But regardless of the nature of the research, an increase in the precision required is almost always associated with the need for more experimental resources. For any given research objective or experimental design, more replications and/or more treatments will usually result in more precise statistical estimates than would fewer treatments or replications.

In addition to increasing the number of treatments or replications, precision can also be increased by closer supervision and better control during the experimental process. Spotty application of fertilizer by hand broadcasting on grass plots can result in large experimental errors as can carelessness and lack of thoroughness during harvest. More time, workers with more skill, or specialized machines or equipment can reduce this source of experimental error, but all will add to the requirements for experimental resources. From a practical point of view, the least expensive means of reducing experimental error from this source is closer supervision by the researcher during all phases of the experimental process. Time spent at the experimental site by the researcher can be highly productive particularly when it is possible to prevent the complete loss of data through carelessness.

The use of prior information can reduce the need for resources by reducing the range and number of treatments otherwise required, by taking advantage of known estimates of variance to minimize the number of replications, and perhaps even by providing all the data needed making further experimentation unnecessary. In the following section we will discuss in more detail the use and utility of secondary experimental data, which is one source of prior information available to the researcher.

The time limit within which a decision must be made is also an important factor in determining the size and complexity of an experiment. If time is not a critical factor, an exploratory experiment of relatively simple design can be conducted and this can be followed by an experiment which is more exact in its focus. The range over which the experiment is conducted can be successively narrowed until the required accuracy is achieved. None of the experiments in this sequence need be large nor complex. But if timing is critical and decisions must be made rapidly, it may be necessary to increase the size and complexity of the experiment in order to include several links of the chain all at the same time. That is, to conserve on time (in this case a very limited resource) other available resources may need to be substituted in order to achieve a useful research product within the time period allowed for making a decision.

Secondary Experimental Data

When experimentation is undertaken in the execution of the research project, the experimental design can and should be tailored to the needs of the project. The data obtained will then be the best available and will be well suited to the needs of the researcher. However, it is not always necessary to conduct an experiment to have suitable or adaptable experimental data for analysis, because in most places where applied research is being conducted, at least some prior experimental data are available. These data, though generated for other research purposes, frequently provide an insight into the nature of the relationships to be studied in the current project, and may provide sufficient data so that additional experimentation need not be undertaken.

In situations where time is a limiting factor, the use of secondary experimental data may be advantageous or even essential, but data generated by another person or from one or more other experiments must be used with caution to assure that they are relevant and comparable. Some adjustment or selection may well be necessary to adapt the data to the current study.

Several means can be used to adapt, select, or adjust data of this nature. One means is to select only the relevant portion of the data and exclude those parts not related to the current analysis. An example would be to select data in a fertilizer experiment from those parcels where potash was at a constant level, eliminate plots where trace elements were included, and make the analysis for the portion of the data in which only nitrogen and phosphorus were variable.

A second method of using secondary data is to analyze the results of several experiments and search for consistencies which may indicate relationships not otherwise evident. In a series of corn fertilization trials in the Cauca Valley, Colombia, production functions resulted, individually, in very poor statistical estimates. For any one trial, little confidence could be placed in the conclusions. But after dividing the trials into soil types, it was noted that there was a great deal of consistency in the functions within a soil type. The form of the curves was similar with each reaching a maximum

41

within a short range, and with calculated optimum applications at about the same levels. Hence, by using information from all the curves, together, general recommendations could be made even though the individual analyses yielded little information.

A third useful method of analysis is to consider the possibility of combinations of data from two or more different experiments which were conducted under similar conditions. This method was used to estimate the optimum stocking rate for fattening steers on grass from three different experiments [50]. Each experiment was conducted to determine the effect of hormones, and all were conducted on similar grasses, under similar conditions and with comparable cattle. Because the rate of stocking was different in each case, it was possible to use a mathematical response relationship from which the optimum stocking rate could be calculated.

One additional point with respect to secondary data should be mentioned. When one has spent time attempting to use secondary experimental data, he appreciates the productivity of any efforts made by the first researcher to preserve the data for other users. Nothing is more frustrating than to discover a description of an experiment that should have provided usable information and then to find the data in such poor shape that they are impossible to use. Simple notations such as units of measurement are often not even included. The thoughtful researcher will leave a clear record of his data so that users who follow will have no trouble interpreting them.

Multi-purpose Experimentation

As a practical matter, a great deal of experimentation is carried out with an orientation that is only partly research centered. An important example in agriculture is the demonstration trial usually conducted by, or in cooperation with, the extension service. One of the purposes of this type of research is to demonstrate the results of research under real conditions, i.e., under conditions which will be applied by the client toward whom it is focused. For this reason rather poor experimental control is to be expected, and accordingly, the experimental design is nearly always quite simple. In many cases, as few as two treatments, with and without a particular input or package of inputs, are included. At times, a complete experiment is conducted at one location; in other trials, different locations are considered to be different replications of the same experiment with all treatments being repeated at each; and in some cases only one or a few treatments are conducted at each location.

It should be obvious that as more and more locations are included in a demonstration trial, exposure to the clients is increased, but experimental control is decreased. Also, with more treatments in a project, more information is possible, but also, the supervision of the project becomes more difficult. Hence, the persons responsible for the project must determine, based on the orientation of the project and the available resources, what the size and scope should be.

Presenting an example of a fairly successfull demonstration trial may be the best means of discussing some of the more important aspects to be considered when inititating a project of this nature. This project was conducted in southwestern Colombia in an area not far from the city of Cali. The area comprised that of a large number of descendants of former slaves who had divided their holdings into an average of about 2 hectares per family. The farms were nearly all in old stands of cocoa, coffee, and plantain which were in such poor condition that monthly income per farm barely reached fifteen dollars. General nutritional deficiency was evident, as was the nearly complete absence of any source of animal protein. Physical conditions for any type of livestock were very bad, and, of course, little of no money was available for supplemental feed.

As part of a development program for this area, it was thought that Khaki-Campbell ducks (an egg laying breed) could prove to be a potential source of animal protein to supplement the diets of these poor rural families. In designing the project, several alternatives were possible. The simplest of those considered reasonable was to select a small number of families (those most apt to be conscientious in maintaining the necessary records), to give each about 10 ducks and provide them wtih a recommended concentrate ration. The results could probably be established with a fairly high level of confidence and would show whether or not the ducks could survive under the conditions of the area. It would also be possible to determine the cost of the eggs to the families and whether or not they would eat the eggs.

A second approach that was considered was to use more families and experiment with several rations for the ducks. At first, three rations were considered adequate, with three families provided with each ration. In this manner, it was hoped to be able to determine not only if the ducks would survive and lay under the severe conditions prevalent in the project area, but also what the lowest cost ration would be. In other words, there would be three different rations to choose from. Although more information could be obtained, more families would be needed and more supervision required to assure that the rations were correctly administered. The extension personnel who were cooperating in the project felt they could find nine families and had sufficient resources to do the majority of the supervision. Over 100 young female ducks were going to be available, so the extension people began to locate the cooperating families.

As it turned out, 16 families were eager to cooperate in the project, and the extension people felt this number should not be too great a supervisory burden. The use of this many families would reduce the number of ducks available per family to seven. It was decided that this would be satisfactory, but that it would be more efficient in the use of the families to include four rations and have four replications of each. With four rations it was hoped that there would be sufficient information to estimate a production function for eggs and with this information be able to determine the lowest cost ration even if it were not one included in the experiment.

The final design included one group with a complete concentrate ration, one with ⅔ of this amount, one with ⅓ of a ration, and a control group with no concentrate. Except for those receiving a full ration, all could receive whatever scraps were available and be allowed to graze (or be fed chopped grass). Material for constructing adequate housing for the ducks was available locally at no cost and all families were required to be ready to receive the ducks by a fixed date.

A one day short course was held with all families participating sufficiently far enough in advance of the delivery date to allow time for construction of the housing. The care of the ducks was discussed and also the design of the experiment and its purpose. The families were all given very simple record forms and instructed in their use (to be able to use the forms it was not necessary to be able to read or write or even count). Cups were provided to measure the exact amount of feed for each duck for each ration so that in the event of losses, the adjustment of the ration would be simple. At the close of the course a meal including duck eggs was served to assure the families that the eggs were good to eat since there was some local bias against eating duck eggs.

With the encouragement of the extension agent, all participating families were ready on the date that the twelve week old ducks were delivered. Initially, extension field men and the researchers checked frequently with the families to ascertain if everything was being done properly. The records were collected every two weeks during the whole project to assure they were kept current.

To be sure, there were rather large differences in the results for each ration. One family, for example, never was rewarded with even one egg. Although they were a bit embarrassed and the neighbors were sure they were not taking good care of their ducks, they took it good naturedly because they were assured that this was not unexpected and was a necessary part of the experiment. Nevertheless, by having four replications and using the average for each ration, adequate information was obtained. It was possible to determine the best level of concentrate to use if the eggs were going to be sold, as well as the ration which resulted in the lowest cost for the eggs if they were to be used to supplement the family diet. The second use, of course, was the primary purpose of the project.

Upon completing the analysis of the results, a field day was held for the families at which time the results were presented. After the general results were shown, those of each family were discussed in an attempt to determine why some had better results (hence lower costs) than others with the same ration. Following this discussion, all the families participated in the the selection of the final recommendations.

In summary, this project was tailored to fit within the resources available and yet to provide the greatest exposure in the project area. Although experimental control was relatively low, sufficient treatments and replications were included to maintain adequate statistical reliability in part due to the presence of a high degree of supervision by the researchers.

Multidisciplinary Experimentation

Multidisciplinary experimentation is another means of conserving scarce research resources. When researchers from two or more disciplines cooperate in a project it is frequently possible to obtain answers for each with little change in the basic experimental design.[3] Too often a researcher in one discipline minimizes the effects of those factors commonly included by others, so that in the absence of cooperation, the product is of low value to other researchers. An example occurs in beef or dairy projects where the researchers raise their own corn for silage without the cooperation of the corn researchers who may be at the same experiment station. The animal researchers may be able to document accurately the effect of the silage on the animals but be unable to estimate animal production per hectare of corn because they had little interest in the corn production. Even more difficulties will arise if it is later desired to make an economic evaluation of feeding silage to animals [46].

Of course, it is not always possible to obtain answers for more than one discipline without increasing the size and the complexity of the experiment beyond manageability and available resources. But it is precisely in those cases when resources are most scarce that it is important to consider the advantages of multidisciplinary research. Most experiments are costly both in terms of time and of other resources so it is obvious that if the results can serve two or three researchers, the additional effort required in planning the experiment can be extremely valuable [60].

---

[3]For a brief and excellent discussion of multidisciplinary cooperation in extending experimental results to practice see [67].

45

Summary

Experimentation and experimental design usually are associated with objectivity, precision, and scientific purity — concepts that imply rigidity and inflexibility in thought and procedure. In basic research this is mostly true, but in applied research, considerations other than pure scientific objectivity can become more important in determining the type of design to be used for any given experiment. Resource limitations will usually force a reduction in the accuracy obtainable, and factors such as demonstration uses will modify the nature of the design ultimately selected. It is important that the applied researcher maintain a flexible attitude with respect to experimentation and the experimental design in order to increase his effectiveness and make him more efficient in his work.

CHAPTER V

NON-EXPERIMENTAL DATA COLLECTION

A major difference between experimental and non-experimental research is the degree of control the researcher exercises over the variables being studied or measured. In an experiment, the researcher controls the design and levels of certain variables and the measurement of phenomena resulting from the experiment. In non-experimental research, the researcher in most instances cannot determine the design and level of the variables nor directly measure the phenomena, but controls only the technique used in measurement (primarily a sample survey and questionnaire). For non-experimental observations sample survey design plays a role comparable to that of experimental design when experiments provide the observations to be analyzed. Again, a competent statistician, if available, can be a helpful consultant.

The non-experimental researcher relies upon interviews and questionnaires to communicate his measurement need to a respondent. Often it is the respondent, not the interviewer, who performs the measurement based upon his experience. The respondent's experience may be compiled, for example, in farm records or consumer budgets, but frequently his responses are based upon a subjective evaluation of the phenomenon as he best remembers it.

Thus, in non-experimental research the researcher usually controls only the general levels of variables, through stratification and sample selection of respondents, questionnaire development and interview training. Successful non-experimental measurement rests primarily upon all of these activities. But even with a good sample, a tested questionnaire, and a well trained interviewer, the researcher cannot completely control interviewer-respondent communication, part of which may be misleading [16].

This chapter will first focus on the selection of respondents and the design and implementation of questionnaires and then discuss verification and preparation of both primary and secondary data for analysis. No specific emphasis is given to unstructured interviews and case studies because most of the approaches included are applicable in these situations. Time and resource restrictions along with the needs expressed by the hypotheses and objectives will dictate whether the measurement process will be a large survey or a case study [31].

Selecting Respondents

The selection of the respondents, or sample design, is an extremely important part of the process of non-experimental data collection.

Because the researcher has no control over the distribution of the factors he is studying (these exist as a result of nature and social organization), the proper selection of the respondents is necessary to assure that the information is being obtained from an appropriate population. Once respondents have been selected and the interview process is completed, a faulty or ill-designed sampling procedure may generate data that do not describe the target population. To resolve this difficulty more time and resources, if available, must be expended than would have been necessary with a more carefully designed sample.

Two commonly employed sampling approaches are the random sample and the stratified sample [8, 18, 20]. For information of a broad census nature, a completely random sample gives each person in the total population (all farmers, high school students, married women, etc.) an equal chance of being chosen. In this manner, results can be associated with the characteristics of that population. Stratified samples where particular sub-groups are chosen are used to assure that each sub-group is equally represented, or to assure uniformity or representation over specific ranges of certain group traits (different farm size or different income groups for example). Such a stratified sample can approximate experimental measurement but with less precision than is normally associated with experimental research. A stratified sample is often more efficient than a random sample in use of scarce research resources because the sample size required may be smaller thus reducing both interview and data processing expenses.

In drawing a sample, one hopes to have available a reasonably good listing of the general population including some descriptive characteristics to aid in stratification. To help determine the sample size, measures of variance within the population may be calculated and used along with resource limitations and levels of confidence desired by the researcher and his client.

Many times in applied research, these general characteristics of the population are unknown, making sample selection more difficult. One researcher studying farms in a jungle area was forced to utilize somewhat unorthodox but otherwise relatively functional techniques [42]. In one of the three zones of the region to be studied maps were available showing farm locations. A simple random sample of farms was easily obtained from these maps. In another zone 'trail maps' were available but the farms were not included. The approximate number of farms in total was known and the trails and farms were, in general, uniformly distributed throughout the region. A complete enumeration to develop the sample frame would have been difficult and would have required more time than was available. The researcher chose to send out interviewers by mule with instructions to obtain an interview at every fifth farm. The

48

sampling problem was thus handled as the interviews were taken. In the third zone, maps were extremely poor and little was available to guide the sampling process. To assure that the entire zone would be represented, the interviewers were instructed to ride their mules for one hour and then take an interview at the nearest farm, again proceeding along the trail. As one might expect, errors did result, but adjustments in the analysis were based upon general observations made by the researcher while in the zone.

Had logistical problems not been so extreme, the above researcher could have used a sequential sampling technique. That is, based upon a predetermined level of confidence, he could have measured the variance in selected population characteristics by using the interview data taken daily while in the field. From this calculation he could then determine when the sample had reached the size needed by reducing the variance to meet the desired confidence level. When that level had been achieved he could have moved to another zone.

A further need for flexibility in the sampling process to allow for obtaining needed information is illustrated by a research project recently completed in the East Uttar Pradesh of India [58]. Initial interview attempts with Indian farmers concerning their land holdings were met with considerable suspicion and apprehension, possibly because of speculation about land ceiling legislation. Individuals conducting planned interviews in villages encountered farmers unwilling to provide the desired information. Yet when the farmers were selected at random while traveling along the roads in the district they responded readily to questions about village and other matters related to land ownership. Since personal identity and specific location of the respondents were not imperative the sampling approach was changed substantially but to the benefit of the entire research effort.

In summary, a flexible researcher who places major emphasis on problems and views his sampling techniques as tools and not ends in themselves will be able to perform applied non-experimental research under much less than optimum conditions.

Designing the Questionnaire

At the risk of seeming elementary, it is advisable to emphasize that questionnaire development must be closely tied to the problem, hypotheses, objectives, analytical techniques, and available resources for the proposed research. Just as the research proposal or plan must reflect a relevant problem, and the hypotheses and objectives must pertain directly thereto, the respondents to be interviewed and the questions to be asked must be relevant to the research proposal. Beyond relevancy, a questionnaire must include only those questions that are highest in priority for analysis and testing. A multitude of questions could be "relevant" or "interesting" but only a limited number can be used fully in the project because of research resource limitations. Interview time, for example, represents a costly resource allocation not only for the research project, but also for the respondent. The researcher must always remember that excessive and/or irrelevant questions try the respondent's patience, reduce interviewer-respondent rapport, threaten credibility of the responses, and waste scarce resources of the research project.

Difficulties in Interpretation and Communication

The method of obtaining information in a questionnaire is the spoken word, so language problems must be given special consideration. Obvious problems arise when questions are translated from one language to another, but even within the same language, word and expression usage vary with different cultural, social, educational, and economic classes. Differences of interpretation may exist between the researcher and the interviewer as well as between the interviewer and the respondent. There will also be differences between respondents in the interpretation of some questions. The magnitude of this difference will increase as the heterogeneity of the sample increases.

The complete elimination of language problems is nearly impossible, but careful consideration of cross-cultural communication difficulties will minimize problems of interpretation in the questionnaire. Two major cross-cultural communication problems in questionnaire formulation are differences in terminology and cultural differences in beliefs and values.

The names by which things are known or the terms which are used to describe something vary widely even within a language and culture group [17]. These differences in terminology must be considered in the construction of the questions. Generally, the lower the general level of education and the greater the degree of isolation, the more important it is to take local usage into account. When groups are isolated, indigenous objects and phenomena frequently have names which are known only within the groups. Objects or phenomena which are not indigenous will probably have similar names, but if they are unknown within the group, they may be difficult to explain. Terms used by professionals may not be understood at all by the respondent, yet if the interviewers are allowed freedom in explaining these unknown terms, an interpretational bias may arise. In some cases, professionals may be able to communicate with each other and with the interviewers with one question, but several questions may be necessary to obtain information from the respondents of a survey.

A common example of varied usage in Colombia is the problem with land measures. Four measures, the **hectare**, the **plaza**, the **cuadra**, and the **fanegada**, are used in different parts of the country. When a survey is conducted, one must construct the questionnaire to account for this variation or risk interpretational errors in measuring farm size, crop yields, and other factors associated with land area.

A different problem arose in Colombia with respect to a question concerning tapeworms in swine.[1] In each of three different departments (states), a different local term had to be used in the question to avoid erroneous conclusions which could have been drawn when determining the respondents' knowledge of the incidence of this parasite in the survey area.

Unknown or misunderstood differences in beliefs and values are the other important cause of interpretational problems. An example can be drawn from interviews with housewives in a Colombian village where the drinking water was badly contaminated [27]. Many of these housewives realized the importance of boiling their drinking water and understood the technique involved but still did not do it. Informal interviews determined how many housewives did boil water, how many realized they should, and how many understood the technique, but the survey failed to determine why the majority did not actually boil water. The difficulty developed because appropriate questions were not asked due to inadequate cultural knowledge. The professionals assumed that boiling water was "good", and based upon survey results and their own normative considerations, they determined the housewives "should" have been boiling water. Only after further questioning did the researcher discover that the villagers believed that water was boiled only for the very ill. Those who drank boiled water were considered to be ill, so to avoid this stigma, many who otherwise might have boiled their water failed to do so.

Designing for Data Retrieval

A questionnaire must lend itself to efficient and useful data collection, processing and analysis. The allocation of time and resources between field collection and data retrieval and verification from completed questionnaires needs careful consideration when designing a questionnaire. Long and detailed qualitative responses to open-ended questions may be more informative than quick semi-quantitative responses involving selection of alternatives, but the time required to correctly obtain, retrieve and analyze one long

---

[1]The question is from a questionnaire which provided information for a Masters thesis [49].

51

response may be the same as for twenty short selective response questions. One must determine which of these two techniques (or combinations of both) most efficiently provide information while minimizing interpretation and response errors.

Both collection and verification are facilitated by logic and consistency and by organizing the questions to lead the respondent with ease through the questionnaire in a manner that stimulates spontaneity in his responses. If one question asks about his marketing problems for rice, and is followed by the age of each member of his family, and the next returns to the price of rice, the interview becomes burdensome and less effective. One exception to this rule is important. If one desires to cross-check or verify a particular response because of interpretational problems or a reluctance to respond accurately, it may be desirable to ask the same question twice, each time in a different manner and at different points in the interview.

Dividing the questions into major groups will help reduce respondent fatigue while maintaining interest. Groups of questions that involve sensitive issues; such as income or profit that may be subject to taxes, might follow several relatively less sensitive groups of questions in order to establish rapport and confidence. Sensitive issues covered prematurely in a questionnaire may destroy the entire interview and thus should be asked toward the close of the interview.

Coding, or preparing the data for computer analysis, is necessary if the data are to be punched on cards for electronic data processing. Precoding which refers to specifying on the questionnaire the column requirements for computer cards, can eliminate the need to transfer the data to code sheets by permitting the data cards to be punched directly from the questionnaire. This method requires more careful editing of the questionnaire but usually reduces retrieval costs and the chance of human error by removing one transfer of data. To be more efficient, however, the researcher must be certain beforehand of how he wants to organize and process the data. The editing and checking process, on the other hand, helps assure that all necessary data will evolve from the questionnaire. Precoding is less advisable for highly qualitative responses or when the researcher has no indication of the range of responses he might expect for each question.

Pretesting the Questionnaire

Pretesting the questionnaire or checking to see if it will obtain the information sought, must be accomplished under actual field conditions before beginning the general survey. No amount of intellectual exercise in the office can substitute for properly testing a questionnaire among the respondents in the area to be surveyed. This phase will determine the weaknesses in the questionnaire, establish whether the information sought can be obtained in a useful form, and also may provide additional information which can help improve the relevancy of the questionnaire.

Whenever possible, the interviewers who will be conducting the survey and the director of the research project should be involved in the pretest to assess problems with the questionnaire and with respondents. Prior to pretesting, interviewers should become familiar with the research project and the questionniare. Each interviewer should be carefully briefed and sent to several of the different areas which have been selected to assure experience with a variety of respondent types. When each interviewer returns, a debriefing with the researcher should be performed immediately. This debriefing should include a review of the completed questionnaire to identify points of misunderstanding, unnecessary repetition, time difficulties and so on. General reactions by both the respondents and the interviewers concerning the subject matter and length of the questionnaire should be discussed.

For some interviews care should be taken in selecting the areas where the pretest will be performed to avoid the danger of contaminating a major target area by pre-testing the questionnaire within its bounds. Especially in close-knit rural areas, advance knowledge of attitudinal responses in the final survey could be affected by word of mouth discussion among potential respondents and those participating in the pretest.

Size of Pretest

The required number of pretest questionnaires depends upon the research problem, the homogeneity of the survey population, the data collection and analytical techniques to be employed, the total number and complexity of the questions to be asked, and the research resources available for the project. The range may be from five to one-hundred or more questionnaires. If the population is very heterogeneous with respect to information required, more interviews will be necessary for an adequate pretest than for a questionnaire to be administered to a more homogeneous population.

If the pretest is also to provide desired information about variance to determine the population homogeneity for sampling purposes, more pretest interviews may be needed than when only a test of the questionnaire is desired. For potentially large studies where general population parameters are unavailable, a rather extensive pretest can aid in designing the sample. The marginal costs associated with these additional pretest questionnaires must be measured against costs associated with sampling errors due to having too few respondents. Extra costs resulting from obtaining excessive questionnaires in the main interview effort may be greater than the costs for a few additional questionnaires in the pretest.

Information Checking

How a pretest improved a particular question in a survey of settlers in the colonization project in Caqueta, Colombia is illustrated as follows [48].

Original question:
Where did you live before you moved here?
Improved questions:
In which Department were you born?
Where did you live before moving to Caqueta?
Where did you live before moving to this farm?

The researcher wanted to know from which departments the settlers had migrated to Caqueta, but the original question only revealed information about the respondent's last move which may have been within Caqueta and not from another department. Even though questions were added, the results were easier to interpret. Based on this researcher's information needs, the first question had no value so he was forced to choose between no question and three questions.

A pretest will assist the researcher to develop an efficient but complete questionnaire. A poorly pretested set of questions designed to obtain information about farm management practices and production costs for potato producers in Colombia illustrates this point.[2] The respondents were asked the number of hectares devoted to potato production for the most recent year at three different points in the questionnaire. At the beginning, producers were asked to give total farm size and land use by crops, pasture, and other classifications. Later in the interview they reported on potato plantings for each **semester**. This information often did not agree with the first question because of double counting. It was difficult to verify the response because plantings in the first **semester** were usually much larger than in the second, so the extent of the double counting was impossible to determine. And finally, when queried about hectares harvested the area given was usually slightly different from that previously used. Could it be assumed that those hectares not harvested were lost to blight or drought? Perhaps, but it would be better to be certain that the difference was not in part due to either intentional or unintentional reporting error. Not only was the duplication unnecessary, but the questions were am-

---

[2]From a portion of a producer questionnaire used in Colombia and summarized in [37].

biguous. A better pretest could have minimized this discrepancy by guiding the preparation of one comprehensive and easily understood set of questions related to hectares of potatoes seeded and harvested for the past year.

After completing the pretest, responses should be tabulated and critically reviewed to determine whether the questions were understood clearly and whether the information which resulted would help to resolve the problem. This step frequently uncovers gaps in information, or responses which are not in the best form for the analyses which are to be undertaken. While the pretest may identify needed modifications in the questionnaire, it also provides an opportunity to determine whether or not the responses are in a form such that the hypotheses can be tested. The researcher should carefully consider what is being collected in the pretest and how it can be used in the hypothesis testing and analysis phases of the research.

Some research advisors insist that their students use pretest data on a practice run through the analysis which slows the research process, but only momentarily. Analyzing pretest data can aid greatly in refining and even specifying the exact analytical tools which typically are not given too much thought until all of the data are collected. This may also provide some ideas as to the format for the final publication. If these benefits are gained from analyzing pretest data then the saving of time at a later stage in the research process will more than compensate for the early time loss.

Often a pretest indicates that either too little or too much information is being collected. At times, the pretest may show that a certain type of information will not be available, and therefore, some modification will need to be made in analyses, hypotheses, or objectives. More commonly, the questionnaire may be gathering more information than is necessary which suggests that costs of getting the excess should be carefully considered relative to potential uses of the information.

Only when pretests have shown that interpretational problems are minimal and that the information can be obtained in the form required, should the questionnaire be finalized and the complete survey undertaken.

Time Difficulties

Because timing is so important, the pretest should be used to help specify interview time problems; time of year, month, week, and day must be carefully considered to fit the respondents' work schedule.[3]

---

[3]Mail questionnaires help resolve time problems but often limit response quality and, due to low response rates, require large samples. A bibliography concerning mail-questionnaire research is found in [42].

If, for example, one is proposing to interview truckers at check stations or consumers at a retail market the time element may be more critical than in interviews with producers. Retailers usually prefer granting interviews when the fewest customers are in the store. For small retailers, a daytime interview can be three or four times as long as an uninterrupted nighttime interview; but if an appointment cannot be obtained for non-working hours, as is often true, the more lengthy daytime interview may be necessary. Agricultural producers probably will find harvest time to be inconvenient.

Numerous examples can be cited but few guidelines can be specified for timing interviews because each survey and often each interview represents a different situation. In any case, should the respondent feel that the questionnaire is completely irrelevant and a waste of time, the responses will lack credibility even when the interviewer is fortunate enough to complete the interview.

Selecting and Training Interviewers

A well-designed and pretested questionnaire will not yield good results if administered by an interviewer who is poorly trained or who has a disagreeable, non-cooperative and culturally or class biased attitude toward the respondent. The contact made with the respondent will not only influence the interview but also influence the respondent's evaluation of the entity for whom the interviewer is working, and his acceptance of action programs that might evolve from the research recommendations. Requisites of a good interviewer include: 1) an interest in and an understanding of the research project, 2) an interest in people and the ability to communicate that interest and sincerity to a respondent, 3) a willingness and ability to follow instructions and definitions without regard for personal beliefs or convictions, and 4) in general, a good public relations attitude.

Training is necessary to familiarize both experienced and inexperienced interviewers with the problem to be studied, the objectives of the research project and the organization sponsoring the work, the questionnaires, the respondent selection procedures and the interviewing techniques. An interviewer's manner of asking questions must be objective and he must be neutral and honest in recording responses. The most important goal of the survey is to get an unbiased opinion or bit of data for each question. Whatever the interviewer thinks of the respondent and his opinions should not influence interviewer objectivity.

Interviewers should be instructed carefully on methods for introducing the research project to the respondent. A most difficult question to answer is a respondent's inquiry about the legitimacy and legality of the study. Letters of introduction and any other documents from official and credible sources should always be in the possession of interviewers. To establish respondent confidence, the introduction used by the interviewer and the letter itself should explain that the interview is confidential, and that the information sought will be used to determine attitudes and characteristics of groups for comparative purposes, and not to illustrate characteristics of specific respondents.

Although the interview must be objective, the interviewer must be given flexibility to obtain a useful response. He should record specific comments and not vague or meaningless generalities such as "because it is interesting" or "I like it because it is good," Why is it interesting? Why is it good? The "I don't know" response is one of the most difficult to manage. One is not sure whether the respondent really did not know, whether he did not understand the question, or whether he did not want to respond for various reasons. The respondent should be encouraged by the interviewer to believe that an "I don't know" response is not an admission of ignorance. A simple guess or an estimate calculated to suit what the respondent feels the interviewer wants is unacceptable. The respondent, however, should not be encouraged to give an "I don't know" response when he most likely can give a meaningful and honest reply.

Verifying Primary Data

Verification of the data after it is collected is necessary before data processing and analysis begin. The researcher must become fully aware of the limits and potential uses of his data based upon the objectives of his research. A careful review of data with well performed revisions at this stage of the research process will reduce the chance of costly errors.

Numerous and detailed verification activities are necessary so the researcher can be confident that the data best measure the phenomena needed to fulfill the research objectives, given the resource limitations placed upon the project. The verification and preparation process must consider techniques for handling missing observations, falsified data, completely inaccurate measures, and so on, as well as the necessary conversions where measuring units have differed. Often the required transformations, corrections, and general manipulations performed on data are dictated by the type of facilities available for processing. Computer processing is not necessary for all research, particularly when projects are small and calculators are readily available. Both time and monetary costs for each alternative should be considered along with the type of analysis desired and the degree of accuracy needed. But regardless of the processing facilities and techniques to be employed, well specified methods for verifying and coding data will improve efficiency in research resource use.

During field data collection and while verifying and coding results, the researcher should accurately record his collection and verification techniques. This record, commonly called a code book, is also oriented to describing data collection and verification techniques and it may also include some data refinement such as frequency distributions, means, standard errors and so on. It not only serves as a guide to his present work, but also serves to indicate in detail what the data represent for use in further research efforts.

Verifying and Using Secondary Data

The researcher usually has no control over secondary data measurement but he has the responsibility for checking very carefully to determine exactly how it was derived, the nature of the aggregations made and if the data will meet the needs of his research [65]. Secondary sources can be classified into two major types: 1) regular and long-run measures of phenomena such as prices, income, production, rainfall, and temperature; and 2) less regular and, in general, short-run measures of phenomena more often associated with another person's research such as crop response to improved inputs, consumer attitudes, and management characteristics.

Time series data can often be difficult to use and verify because they are secondary and not necessarily designed for a specific research project. Because time series frequently cover extended time periods, collection is performed by an agency which is responsible for maintaining continuity in the collection process. No particular person is responsible for collecting these data over the years, but certain procedures are established and revised occasionally to accurately and adequately measure the particular phenomena under study. Nevertheless, definitions should be checked carefully so that the researcher can decide what the data in question are supposed to measure, and since he has no control over the collection of these data, he must satisfy himself that they in fact do measure the phenomena in which he is interested.

Comparability may be the most common difficulty in using time series data. This refers to 1) comparability of different time series which purport to measure the same phenomena and 2) comparability of different periods within the same series. For example, due to plant breeding and food processing programs, the corn produced and marketed fifty years ago and the corn produced and marketed today are virtually two different products. A price series for corn covering this period of time is not comparable during the entire period. Coupled with product changes in a price series is the equally difficult problem of non-comparability in currency values caused by price inflation or deflation in the national economy.

A serious comparability problem, and one that is difficult to detect within a time series, is a change in definition. For example, a change in quality standards will affect a price series, but may not necessarily be specified. A change in the point of measurement such as the location of price determination from the farmer-assembler level to the wholesaler-retailer level will influence the value of the economic variable in question. In census data one may find a change in the definition of a farm. Similarly, changes in the population of a city may in part be due to changes in the boundary lines.

Non-comparability between series attempting to measure the same phenomena is common particularly in developing countries because time series measurement systems have often not been standardized under the authority of one agency. Unfortunately, in many cases these entities neither clearly define the phenomena to be measured nor the precise points and times of measurement. And when the points and times are defined, a change in definition may not be spelled out after it is made.

As an example, consider the separate series of potato production data compiled by the **Caja Agraria** (a public agricultural credit agency) in Colombia and IDEMA (a public agricultural marketing agency) [36, p. 230]. These series are vastly different, with the production data of IDEMA averaging about 40 percent lower than the data of the **Caja Agraria**. This difference is probably due to differences in methods of measurement of the series. The **Caja Agraria** primarily develops their series from producer loan data, while that of IDEMA is based on measures taken in rural markets. Thus, **Caja Agraria** more closely estimates total production and IDEMA measures the production entering commercial channels. If the researcher is interested in the volume of commercial sales of potatoes, the IDEMA data probably are more accurate. On the other hand, if the researcher wishes to estimate per capita potato consumption, including both rural and urban consumers, the IDEMA series would under-estimate this variable by about 40 percent less losses and seed requirements.

Where a time series is needed but unavailable or possibly inappropriate for a specific problem, cross-section data sometimes can be substituted. Cross-section data (experimental and non-experimental) represent phenomena measured only at one point in time or a limited number of times. One means of using cross-section data to simulate longevity is in questionnaire research where respondents recall information concerning an event at present, last year, five years ago, and so on. Care must be observed in interpreting these measures because of what has been termed a telescoping bias. That is, the tendency to completely overemphasize or underemphasize certain phenomena by projecting present conditions to previous points in time while forgetting others that may have influenced these observations.

Summary

The success of non-experimental data collection rests upon the ability of the researcher to accurately sample the defined population and, once the sample is drawn, communicate with the selected respondents. In designing the questionnaire which is the means of obtaining information from the respondents, two important communication problems must be considered: 1) differences in terminology between various groups, and 2) cultural differences in beliefs and values. Pretesting the questionnaire under actual field conditions can provide information both about its effectiveness as a data gathering tool and information on the population which can help in establishing sample size.

As in the case of experimentation, guidelines for optimum sample selection procedures and questionnaire design in non-experimental research can be stated, but in applied research it is important that the researcher remain flexible in his attitude. Scientific perfection can serve as a norm, but the researcher must remember that for the client, it is almost always better to have some information to help him make his decisions than to have no information except that the researcher is still designing a better questionnaire or trying to decide on the best means of choosing the respondents.

CHAPTER VI

DATA UTILIZATION — WHAT DOES IT ALL MEAN?

The real skill of the applied researcher comes into play after the collection of the data has been completed. Experience and imagination have a particularly high payoff in the analysis and the interpretation of the data and can make a difference between a useful project and one which ends up in a file drawer. It is in this process that the researcher finally comes down to the point of determining what the data entail; data do not "speak for themselves" but must be interpreted and analyzed. The researcher must draw conclusions from the analysis and in the end make recommendations to his client to help in resolving the problem that originated the project. This, of course, is the reason for undertaking applied research in the first place. No amount of planning, no elegant data collection procedures, and no sophisticated analyses are going to help the researcher who is too timid when the moment of truth arrives to utilize all his information, draw meaningful conclusions and make appropriate recommendations to the client.[1]

When this moment arrives, the client is expecting a useful product and the researcher is the most knowledgeable person available to him. At the conclusion of this project, the researcher should know more about the subject being studied than anyone else with whom the client has contact. If this is not so, the client should have gone elsewhere for his information. And if the client had not needed the information he would not have contacted the researcher nor utilized the other research resources. Hence, the researcher must assume that his knowledge is vital to the client and that the client desires the fullest utilization of the resources which have been expended by the project. Also, because of the nature of applied research, the researcher is usually facing a deadline, so additional data collection is seldom possible. Conclusions and recommendations must be made on the basis of the data at hand because that is the best information that is or will be available within the allowable time and resource restrictions.

Rather than cover the myriad of analytical procedures available to the researcher, which are presented in detail in a variety of good sources, this chapter treats the more personal aspects involved in the interpretation of research results.[2] These are the aspects which might be called in part, the art of research, and which also might be

---

[2] A sampling of basic texts covering analytical techniques includes [13, 20, 26, 6, 32].

[1] For references particularly concerned with interpretation and communication of research results see [21, 44, 57, 78].

called subjective analysis or, by the purist, personal viewpoints and judgements. We prefer to think of subjective analysis as flexibility in ones attitude toward the scientific procedure and the approach discussed in this book. This is the attitude that allows the researcher to look beyond the numbers which result from the analysis. It is this attitude that allows the researcher to completely milk the data and draw out all the information which might be of help to the client. Furthermore, this attitude encourages the good applied researcher to insist on a role in the interpretation of the statistical or other analyses which have been used (either by him, by another person, or by a computer) rather than accept these impersonal results without question.[3]

Applied research is not useful to the client when the researcher reports that he had to go back for more data so has no conclusions, or that based on such and such a level of confidence there is no significant relationship between the variables. It is more useful to report that, although a relationship is not strong, and high confidence cannot be placed in the conclusion, there is a tendency toward a particular relationship and that because this response is logical, the relationship can be used in resolving the problem. Or perhaps, just as strong a statement could be made for there not being a relationship between certain variables. The client is depending on the researcher to draw a conclusion and make a recommendation. The client, then, armed with this best estimate of the researacher, ultimately makes the decisions to be taken to alleviate the problem.

Because the client must make a decision, it is also necessary that he understand the information which the researcher presents as the results of the research. Too often the investigator writes his report as if he were communicating only with other professionals and thereby ignores the needs of the client toward whom the presentation must be directed.

In this chapter, two factors associated with the utilization of data in applied research are discussed. First, an attitude of flexibility in the analysis of data prevents the researcher from becoming boxed in by tradition to the point that he is unable to understand what his data are trying to tell him. A lack of flexibility on the part of the researcher can impede the complete interpretation and full utilization of the data so that the client does not achieve maximim benefit from his investment in the research undertaking. The second aspect of data utilization to be covered is the presentation of the results in a form such that the client can adequately understand the implications of the project and use the results accordingly in his decision-making process.

---

[3]For a lengthy debate concerning the role of social science research in prediction and prescription see [45, 59].

Flexibility in Interpretation

Flexibility in the interpretation of the data and the analyses of the research project does not mean their manipulation to achieve predetermined results. This defeats the purpose of a project undertaken to resolve a problem. Flexibility refers to the capability to really comprehend what the data and the analyses mean and how the relationships which they express can be used to advantage by the client in making a decision.

In close conformity to the needs and desires of the client, the applied researcher should utilize all his training and experience as well as the knowledge gained from the current project in order to provide useful information. This includes a complete examination of the results to determine their meaning as well as their reliability. Overemphasis on measures of reliability and insistence on rigid standards frequently set for more optimum conditions or different problems, reduce the ability of the researcher to explore the data in more detail and to fully understand the meaning of the results.

Meaning of the Results

A common fault in the research process is to accept results of the analyses as something sacred, even if they do not appear logical. An important aspect in the research process is the selection of choice criteria, the measures of performance, efficiency or success which serve as guides in the theoretical construction of the problem statement. The performance criterion which should have been of concern to the researcher in our opening dialogue of this book was the increase in the total production of certain crops in his country for which trade agreements had just been made. An agronomist working on the development of a new variety may have as a performance criterion the resistance of the crop to a certain disease. The same agronomist also, of course, has a secondary criterion, that of increasing production per unit of land area in which the crop is grown. A farm economist usually considers the maximization of profit to some resource base as the most relevant performance criterion to use.

Blind adherence to a predetermined set of performance criteria and the lack of flexibility in considering alternatives can frequently obscure the real nature of the problem which is being treated. Three examples of errors in interpretation owing to the misuse or the misunderstanding of the performance criteria will be discussed. These examples should provide the researcher with some ideas of the kind of flexibility that is needed in the interpretation of his research results.

Example 1.[4] The performance of potato producers in Colombia, as

---

[4]Taken from [37].

in most places, has been measured by research and extension specialists in tons per hectare simply because of professional tradition. Experimental yields per hectare have improved annually since the early 1950's to the point where they now at least triple and often quadruple average producer yields. Yet yields on both commercial and subsistence farms have risen only slightly over the same period. The research and extension programs are often under criticism for not stimulating at least part of the yield increases which are known to be possible. At the same time farm research reveals that both large and small producers are appying fertilizer and pesticides at levels near to those recommended, and many are also using improved seed. Total potato production has increased to keep pace with population growth mainly through the dedication of more land to the crop, but yields per hectare remain embarrassingly low. Should producers, agronomists, and extension specialists be embarrassed? Maybe not.

Except for the few potato farmers who rent land for potatoes, land represents a relatively low cost input to even the small producer. The small farmer in the potato regions usually produces only this one crop as a cash crop. Hence, he considers all his good land as potential potato land, and if feasible, he usually has marginal land that can also be put into potatoes. The farmers use as much of their best available land as necessary to consume their seed supply. It is this seed supply which is an expensive input in the eyes of the producer. This is because during the year he has the alternatives of selling or eating the potatoes which are held in storage, and some of which he is saving for seed. As the time for seeding gets nearer, the price normally rises and the potatoes tend to spoil, creating even more temptation to sell the small amount of seed stock remaining. The purchase of seed, if necessary, creates a real hardship on him and his family. Hence, at planting time a producer has only a limited supply of good seed available. As a result, even if he has a small farm, he may have more land available than seed to plant on it.

Thus, it is not yield per hectare which most interests the potato producer but yield per amount of seed used. He even reports his yields in a yield-seed ratio such as 20:1. This goal (the farmer's performance criterion) implies wider spacing between rows and between plants than is recommended by agronomists who consider yield per hectare as a performance criterion. The result is a higher product-seed ratio, but a relatively low product-land ratio. Nevertheless, the potato producer is reacting rationally to his resource situation and neither he nor his scientific advisor should be embarrassed by low per hectare yields. In fact, the product-seed ratio has improved substantially in large part because of the technical assistance the producer is receiving. But at the same time, many research and extension workers remain frustrated because they are relying on an inappropriate, though traditional, performance criterion.

Example 2.[5] A similar situation has occurred in West Pakistan where irrigation is necessary for crop production. Historically, water, whether provided by canals or by Persian wheel wells, has been a much more limiting resource than land even though the average farm size in many areas is only about seven acres. Contrary to recommendations, farmers were not using sufficient quantities of irrigation water to obtain highest yields per acre. Rather they spread the available water over more land so as to obtain maximum production per unit of water (they still were not able to use all of their land at any one time). Before more water was made available, an increase in water application per acre would have reduced the amount of land they could irrigate and would have reduced the crop-water ratio, resulting in less total crop for the very limited quantity of water. The gain in yield per acre clearly was not an appropriate goal. Of course with more water and the use of new technical inputs, increased production per acre became a relevant goal.

Example 3.[6] Returning to Colombia, some questions have arisen recently concerning productivity measures for dairy farms. Performance of a dairy herd is traditionally measured by the average pounds of milk produced annually per cow for the entire herd including dry cows. The goal of a dairy improvement program is usually to maximize this average. This performance criterion requires that breeding programs develop large cows with well developed udders and other characteristics conducive to high yields per animal. It has also been assumed in Colombia that protein is lacking in the rations for dairy cattle.

Preliminary research shows, however, that the very hilly, mountainous and rainy pasture conditions found in Colombia may require both different nutrition and breeding research programs. The large dairy cow cannot move about in the mud and hills nearly as well as a small one. Maybe three small cows could better adapt to these conditions than two large cows while better utilizing the same amount of pasture even though yield per cow would be sacrificed. Of course milking costs, veterinary costs and so on for alternative herd sizes should be considered. As to nutrition, the energy requirements for the cows appear to exceed most expectations making energy rather than protein the most important deficiency in the usual ration. These preliminary conclusions have resulted because of a flexible at-

---

[5]Experience by Peter Hildebrand on an assignment with Tipton and Kalmbach, Inc., Engineers in West Pakistan, 1964-1966, and reported in [33].

[6]Research notes reported by Rex Rehnberg and L. C. Garrison, Universidad Nacional — Instituto Colombian Agropecuario, Medellin, Colombia.

titude on the part of the researchers interpreting data which initially appeared not to be very logical.

Reliability of Results

The degree of reliability of research results is nearly always assessed by statistical tests of significance which utilize confidence limits or levels to judge how much reliability one may put in the results. A common research practice is to set confidence limits before data collection is undertaken and then to reject results which cannot be measured to that fixed degree of precision. For example, a null hypothesis is accepted if the tests of significance that are used reach a certain level of confidence and rejected if that level is not reached. Terms in an equation fit by regression analysis are accepted at certain levels of significance and rejected below these fixed levels. In analysis of variance, differences in responses to various treatments must meet certain levels of significance before it is said that the differences are real and significant.

For some research projects, however, significance tests may be impossible and even unnecessary depending upon the nature of the problem and resources available to resolve it. McPherson notes, for example, that economists, when adjusting experimental data for predicting a production function, can rely upon the experience of specialists in several basic science disciplines for knowledge about unmeasured environmental and agronomic conditions which influence crop production. "The fact that 'expert's' estimates would not be subject to statistical tests of significance does not mean necessarily that such estimates for particular purposes would be less accurate than estimates derived from sources that lend themselves to statistical tests." [67, p. 804].

In other research projects, strict adherence to predetermined levels of significance is necessary in order to preserve objectivity and protect the client from unnecessary risk. Experimentation with drugs is a common example where very high significance levels must be maintained if the drugs are for human use.

But there is also a great deal of research, particularly applied research, where adherence to the strict observation of high significance levels can unduly restrict the amount of information which is made available to the client, and can, in fact, be detrimental to his best interests. Consider, for example, the case of a low cost practice which appears to increase the yield of a particular crop, but the increase in yield is not significant at the 90 percent or 95 percent levels of confidence. If we tell our client that there was no significant increase in yield, he will probably decide against its use. But if the cost of the practice is low, then the financial risk to the producer is also low (assuming that there is no reduction in yield due to the use of the practice). It might well be that the producer is willing to accept this risk if he has even a 50/50 chance of getting an increase in yield that would be worth 10 times his investment if, in fact, there really was an increase. In this case, wouldn't it be better for the client to receive information saying that there was an increase in yield in most cases with the use of the practice, but it requires a reduction in confidence limit to, say, 70 percent to be a statistically significant increase?

A 70 percent confidence level, in essence, means that if this same experiment were repeated 10 times under the same conditions chances are that at least 7 of the 10 experiments would indicate an increase in yield. Note that neither a 70 percent nor a 95 percent confidence level tells us that there will or won't be an increase in yield on any particular farm. That effect is only implied by conducting the experiment under conditions similar to those of the client or clients. But if there is a valid similarity between the experiment and the conditions under which the client will use the practice, then the costs and the probable returns of the practice to the producer (the economics of the practice) will ultimately determine if the client will utilize that particular practice on his farm.

In applied research, it is important to remember that the client is waiting for conclusions and recommendations. If the research process has been thorough, then no additional information will be available to help the researcher in making a conclusion unless additional research is conducted. If the researcher insists on additional research before he is willing to make any conclusions and recommendations then it will mean a delay before the farmer (client) can put the practice into effect. This represents potential profit foregone by the client (if in fact a change would have been profitable) and should definitely be considered by the researcher and by the client before making a decision not to draw any conclusions from the results of the completed project. When necessary and appropriate, one can utilize sensitivity analysis to determine the economic costs of making the wrong decision [1, pp. 39-45, 54, 66].

Although the decision to repeat the research may mean a delay of only one year to the client, the economic consequences as well as the decision not to draw a conclusion are related to the so called Type I and Type II errors which can be made in accepting and rejecting hypotheses. A hypothesis is stated in such a form that it is either true or false. The research process generates the evidence upon which the researcher will decide whether or not to accept the hypothesis as stated. But any acceptance or rejection of a meaningful hypothesis is always associated with a probability of making an error (it is virtually impossible to be 100 percent sure that a hypothesis is true or that it is false). A Type I error is committed when a hypothesis is rejected (thought by the researcher to be false) when, in fact, it really is true. If there truly is a response to a practice and the research hypothesis states that there is, then the researcher is committing a Type I error if he advised the client that there is no response because his research indicated that the hypothesis was false. Note that the same consequences follow if the researcher, because of adherence to a too high confidence level is unwilling to decide and hence advises the client that he is unable to determine if there is a response.

In this same example, the researcher is committing a Type II error if he accepts the hypothesis (thinking it to be true) when, in fact, it is false. In this case he advises the client that the practice results in an increased yield when, in fact, it does not. The consequences to the client of the two types of error are obviously different.

If the client follows the conclusions (and presumably the recommendations) of the researcher then in the first case (Type I error), the client does not adopt the practice which would have increased yield and foregoes potential income which he would have earned if he had invested in the practice. In the case of a Type II error, the client invests cash in the practice without results and loses the money invested. The Type II error can be even more serious if the practice actually **reduces** yields (the true effect is negative and not zero).

There is an important relationship between the two types of errors such that a change in the confidence level in an attempt to reduce one type of error increases probability of making the other type of error. Too much concern with the Type II error (the one most commonly considered) can raise confidence levels so high that a Type I error becomes highly probable. When a researcher wants to be very sure that there is a positive response before he recommends a practice, he may, in effect, be advising his client that there is no response even though one does exist. This is usually the case when the only recommendation made from an experiment is that it should be repeated with better control, or that a survey must be repeated with a larger sample or better questionnaire.

The need for flexibility in applied research should be obvious from the above example. A researcher may not need to involve the client in decisions about specific tests once the general confidence guidelines have been specified. But a good dialogue throughout the data interpretation process may be important if new questions arise concerning the overall confidence desired. For some problems which may be extremely difficult to research or for specific hypotheses, the client may be willing to accept a recommendation which should hold true only a majority of the time. Other issues may be more critical and require very high levels of confidence. But neither the client nor the researcher should automatically insist on high confidence levels because to do so may result in the loss of a great deal of useful information.

Finally, the end result of an applied research program may **not** with any degree of confidence resolve the problem to which it was directed. Lest this statement seem completely out of context after all that has been said in the book, we should hasten to add that few problems are totally resolved. In many cases, the end result of research is the identification of other related problems and possibly hypotheses concerning their resolution. Often, when reporting applied research results, definitive recommendations which may help resolve the problem are accompanied by a new question or hypothesis. This is true because applied research is a continuous quest for knowledge which appears to have few bounds beyond time and resource restrictions.

Presentation of the Results

By now the need for drawing conclusions from applied research and making recommendations to the client should be obvious to the reader. The researcher is the expert to whom the client is looking for help. If the reseracher is unwilling to fulfill this role, there is little justification for his having undertaken the project.

We have encouraged flexibility and the use of the researcher's experience and his opinions in the interpretation of the data and in the analyses. This is important to the client. But the client is also entitled to a clear exposition as to what part of the conclusions result from the conjecture of the researcher and what part is attributable to the pure analysis of the data. As a scientist, the researcher is obligated to report the results of the research project honestly and objectively. As a professional hired to provide information for resolving problems, he is also obligated to interpret these results in accordance with the needs of the client.

The final stage in the applied research process is the communication of the results to the client. It is as essential that the client comprehend what the researcher has to tell him as it is for the researcher to understand the meaning of the data. The researcher is a scientist whose job it is to understand complicated analyses and confusing data. The client has other interests and ordinarily does not have the same training, so the researcher must report his findings in a language that the client will understand.

In many cases it is useful to the client to have the results presented in more than one form. He is entitled to, and should receive, a complete report on the project including a review of the problem as finally defined, the hypotheses and the objectives. These all appear in the final report much as they were written in the project proposal. In discussing the conclusions, it is necessary to refer to the hypotheses and show which were accepted and which rejected and how each affected the final conclusions. A clear exposition of this part of the research process provides the client with the basis for understanding the recommendations made by the researcher.

The extent to which the researcher should detail the analyses and data collection procedures — the more technical aspects of the research — depends on the familiarity of the client with these techniques and his ability to understand them. Frequently, the researcher can make one technical report and another, briefer report can be written in non-scientific language, excluding the technical details. The detailed report can be used by the client or another professional to verify the validity of the research procedures should the need arise or to serve as a guide to further research, while the shorter report will usually be more functional for the client to use in his decision-making process.

The majority of applied research in agriculture serves farmers as the principal clients through the activities of an extension service.[7] Many farmers are not interested in how the recommendations came to be made but they need to know what the recommendations entail and what they may mean if adopted. The farmer depends on the researcher and the extension specialist or agent to interpret the results of the research process. In this case it is most useful to prepare a research report which presents only the specific recommendations and the nature of the effect or response which can be expected if the recommendations are followed. An excellent procedure is for the researcher to present results in a form directed toward extension specialists and agents and then work with the extension personnel to prepare a non-technical publication for use in the extension programs.

A thesis is a research report whose language and content are directed necessarily toward other researchers — primarily the student's thesis committee. Few theses are usable in their entirety as a research report for more popular distribution, but they can serve as a basis for the generation of a series of documents, each directed toward a different client or audience. Articles for technical journals can report the factors of scientific interest which occurred in the research process. One or more popular reports or bulletins can be directed toward the ultimate clients or intermediate groups such as the extension service. Finally, leaflet type publications which pre-

---

[7]Research, extension and education can and must blend together in addressing current and future problems. The need and approach are documented in many sources including [41, 44, 47, 51, 57, 78].

sent the recommendations in simplified form can also be prepared if the research is applicable to a wider audience.

Regardless of who the client is, the need to present him with the results of applied research cannot be over-emphasized. Too often, especially in institutional research where the client tends to be an impersonal group not known directly by the researcher, the results of completed projects are not prepared in a form which is useful to the client. When this occurs, the client can become disenchanted with the research institute and financial difficulties often result. Hence, one of the best ways to assure a continuing source of research funds and a demand for research services is to present the client with a useful research product.

Summary

The culmination of the applied research project is the presentation of the results to the client. The form in which they are presented will depend on the ability of the client to understand the nature of the research process. Many clients will be concerned only with the recommendations of the researcher and what they can mean to him if he were to follow them. Other clients will desire a complete research report including a review of the problem being studied, the hypotheses and the objectives of the project. The sophisticated client will also want information on the data collecting procedure and the analyses which were used.

Even the sophisticated client depends on the researcher for an interpretation of the results and will want to consider the researcher's recommendations in the decision-making process. At the termination of the project, the researcher should be considered as the expert in the subject and he should use his knowledge and experience accordingly in interpreting the results for the client, who expects the full utilization of the resources which he has invested in the project.

In this book, we have described the process of planning and executing applied research undertaken for the resolution of a specific problem. Planning includes the specification of the problem in such a form that it is researchable within the resource limitations facing the researcher, the formulation of testable hypotheses which suggest meaningful solutions to the problem, and the delineation of the objectives which describe what is expected to be achieved by the project. The execution of the project includes the collection and analysis of data, the drawing of conclusions, and the making of recommendations.

Throughout the book, the justification for applied research is considered from the point of view of service to a client. The client's need to make a decision imposes a deadline on such research and the importance of the problem along with the nature of the research to be conducted, determine the quantity of resources required for the project and available to the researcher. Within these time and resource limitations, the researcher must strive to create a product useful to the client. With this book, we hope we have presented the researcher with an approach to research under these conditions which will help him to become efficient and therefore to be of greater service to his clients.

REFERENCES CITED

Books

1. Ackroff, Russell L. *The Design of Social Research*, The University of Chicago Press. 1953.
2. Arnon, I. *Organization and Administration of Agricultural Research.* Elsevier Publishing Co., Ltd. Amsterdam, London, New York. 1968.
3. Baum, E. L., et. al. (eds.). *Economic and Technical Analysis of Fertilizer Innovations and Resource Use.* Iowa State University Press, Ames. 1957.
4. Baum, E. L., Earl O. Heady and John Blackmore (eds.). *Methodological Procedures in the Economic Analysis of Fertilizer Use Data.* Iowa State University Press, Ames. 1956.
5. Boulding, Kenneth. *The Image.* University of Michigan Press, Ann Arbor. 1956.
6. Brownlee, K. A. *Statistical Theory and Methodology.* John Wiley & Sons, Inc., New York, London, Sydney. 1965.
7. Cochran, Willard G. *Experimental Design.* John Wiley & Sons, Inc., New York. 1967.
8. Cochran, Willard G. *Sampling Techniques.* John Wiley & Sons, Inc., New York, London, Sydney. 1963.
9. Cohen, Morris R. and Ernest Nagel. *An Introduction to Logic and Scientific Method.* Harcourt Brace & Co. 1934.
10. Conant, James Bryant. *On Understanding Science; An Historical Approach.* New York — New American Library. 1951.
11. Conant, James Bryant. *Science and Common Sense.* New Haven, Yale University Press. 1951.
12. Dewey, John. *Logic, The Theory of Inquiry.* New York, H. Holt & Company. 1938.
13. Ezekiel, Mordecai and Karl A. Fox. *Methods of Correlation and Regression Analysis.* John Wiley & Sons, Inc., New York, London. 1963.
14. Friedman, Milton. *Essays in Positive Economics,* Part I. University of Chicago Press. 1953.
15. Gibson, W. L. Jr., R. J. Hildreth and Gene Wunderlich, (eds.). *Methods for Land Economics Research.* University of Nebraska Press, Lincoln. 1966.
16. Goode, William J. and Paul K. Hatt. *Methods in Social Research.* McGraw Hill Book Company, Inc. 1952.
17. Hall, Edward T. *The Silent Language.* A Premier Book published by Doubleday & Company. 1959.

18. Hansen, Morris H., William N. Hurwitz and William G. Madow. *Sample Survey Methods and Theory.* Volumes I & II. John Wiley & Sons, Inc., New York, London. 1953.
19. Kaplan, Abraham. *The Conduct of Inquiry.* Chandler Publishing Co. 1964.
20. Kmenta, Jan. *Elements of Econometrics.* The MacMillan Company, New York. 1971.
21. Mann, Peter H. *Methods of Sociological Enquiry.* Schocken Books, New York. 1968.
22. Miller, Delbert C. *Handbook of Research Design and Social Measurement.* David McKay Company, Inc. 1964.
23. Nagel, Ernest. *The Structure of Science.* Harcourt, Brace and World. 1961.
24. Noltingk, B. E. *The Art of Research, A Guide for the Graduate.* Elsevier Publishing Company, Ltd. Amsterdam, London, New York. 1965.
25. Pearson, Karl. *The Grammar of Science.* London, J. M. Dent & Son, Ltd. 1937.
26. Peng, K. C. *The Design and Analysis of Scientific Experiments.* Addison-Wesley Publishing Co., Reading Massachusetts. 1967.
27. Rogers, Everett M. *Diffusion of Innovation.* The Free Press of Glencoe, and MacMillan, New York, and London. 1964. pp. 7-12.
28. Runes, D. *Dictionary of Philosophy.* Philosophical Library, Inc. 1960.
29. Salter, Leonard A. *A Critical Review of Research in Land Economics.* The University of Wisconsin Press, Madison. 1967.
30. Schuh, G. Edward. *Research on Agricultural Development in Brazil.* The Agricultural Development Council, Inc. New York, N.Y. 1970.
31. Sudman, Seymour. *Reducing the Cost of Surveys.* Chicago Aldine Publishing Co. 1967.
32. Theil, Henri. *Principles of Econometrics.* John Wiley & Sons, Inc., New York, London, Sydney, Toronto. 1971.
33. Tipton and Kalmbach. *Regional Plan, Northern Indus Basin-Development and Use of the Water Resources of the Indus Basin.* Volume II. Economics, West Pakistan Water and Power Development Authority and Tipton and Kalmbach, Inc., Engineers. 1967.
34. Zetterberg, Hans L. *On Theory and Verification in Sociology.* The Bedminister Press. 1965.

Bulletins & Journals

35. Allin, Bushrod W. "Theory: Definition and Purpose", *Journal of Farm Economics*. Vol. XXXI. No. 3. August 1949. pp. 409-417.

36. Andrew, Chris O. "Improving Performance of the Production-Distribution System for Potatoes in Colombia". Ph.D. dissertation, Michigan State University. 1969.

37. Andrew, Chris O. "Problemas en la Modernizacion del Proceso de Produccion de Papa en Colombia". Departamento de Economia Agricola, Instituto Colombiano Agropecuario, Boletin Departamental No. 6. Marzo 1970.

38. Ayer, Harry W. and E. Edward Schuh. "Social Rates of Return and Other Aspects of Agricultural Research: The Case of Cotton Research in Sao Paulo, Brazil", *American Journal of Agricultural Economics*. Vol. 54. No. 4. Part I. November 1972. pp. 557-559.

39. Back, W. B. "Philosophy, Methods and Status of Agricultural Economics", *Journal of Farm Economics*. Vol. XLIII. No. 4. Part I. November 1961. pp. 898-909.

40. Bear, D.V.T. and Daniel Orr. "Logic and Expediency in Economic Theorizing", *The Journal of Political Economy*. Vol. 75. No. 2. April 1967. pp. 188-196.

41. Boehlje, Michael, Vernon Eidman and Odell Walker. "An Approach to Farm Management Education", *American Journal of Agricultural Economics*. Vol. 55. No. 2. May 1973. pp. 192-197.

42. Brandow, G. E. "Methodological Problems in Agricultural Policy Research", *Journal of Farm Economics*. Vol. XXXVII. No. 5. December 1955. pp. 1316-1324.

43. Buse, R. C. "Increasing Response Rates in Mailed Questionnaires", *American Journal of Economics*. Vol. 55. No. 3. August 1973. pp. 503-508.

44. Caunder, Ray. "Identifying Extension's Community Resource Development Clientele and Adapting Economic Information to the Needs of Each", *Southern Journal of Agricultural Economics*. Vol. 1 No. 1. December 1969. pp. 109-112.

45. Christensen, Robert L. "Logic, Science, and Economics", *Journal of Farm Economics*. Vol. 48. No. 1. February 1966. pp. 137-139.

46. Escobar, Gustavo, et. al. "La Ceba de Novillos en Confinamiento". Departamento de Economia Agricola, Instituto Colombiano Agropecuario. Informe No. 6. Noviembre 1971.

47. Farrell, Kenneth R. "New Priorities in Agricultural Research: Implications for Economists in the West — An Extension

View", *Western Agricultural Economics Association*. Proceedings, 44th Annual Meeting, 1971. pp. 10-12.

48. Feaster, Gerald. "An Analysis of the Relationship Between Infrastructure and Agricultural Development in Caqueta, Colombia". Departamento de Economia Agricola, Instituto Colombiano Agropecuario. Boletin Departamental No. 14. December 1970.

49. Gallo, Alonso. "Analisis Economico de los Problemas en la Produccion de Cerdos en Anitoquia, Antiguo Caldas y Valle". Unpublished M.S. Thesis, Universidad Nacional — Instituto Colombiano Agropecuario, Tibaitata, Colombia. 1971.

50. Gallo, Alonso and Hildebrand, Peter. "Analysis Economico de la Ceba de Novillos en Pastereo Rotacional". Departamento de Economia Agricola, Instituto Colombiano Agropecuario. Informe No. 5. Regional No. 5. Septiembre 1971.

51. Guither, Harold D. "Extension Education in Bargaining Among Midwest Livestock and Grain Producers", *American Journal of Agricultural Economics*. Vol. 55. No. 1. February 1973. pp. 58-60.

52. Halter, A. N. and H. H. Jack. "Toward a Philosophy of Science for Agricultural Economic Research", *Journal of Farm Economics*. Vol. XLIII. No. 1. February 1961. pp. 83-95.

53. Halter, A. N. "The Identification of Problems in Agricultural Economics", *Journal of Farm Economics*. Vol. XLII. No. 5. December 1960. pp. 1459-1471.

54. Havlicek, Joseph and James A. Seagraves. "The Cost of the Wrong Decision as a Guide in Production Research", *Journal of Farm Economics*. Vol. XLIV. No. 1. February 1962. pp. 157-168.

55. Hildebrand, Peter E. "Analisis Agroeconomicas Mediante Superficies de Respuesta". Ministerio de Agricultura y Ganaderia, Direccion General de Economia Agricola y Planificacion, Departamento de Administracion Agricola, San Salvador, El Salvador. Septiembre de 1972.

56. Hildreth, R. J. and Castle, E. N. "Identification of Problems", in *Methods for Land Economics Research*, edited by: W. L. Gibson, Jr., R. J. Hildreth, and Gene Wunderlich. University of Nebraska Press, Lincoln. 1966.

57. Holt, John, Charles R. Pugh and William L. Brant. "Educational Programs for Commercial Agriculture and Agribusiness", *Southern Journal of Agricultural Economics*. Vol. 5. No. 1. July 1973. pp. 47-54; and discussion by J. Michael Sprott, pp. 55-57.

58. Jhunjhunwala, Bharat. "Agricultural Mechanization, Rural Income Distribution and Unemployment in Faizabad District, East Uttar Pradesh". Unpublished Ph.D. dissertation, University of Florida. 1973.

59. Kelso, M. M. "A Critical Appraisal of Agricultural Economics in the Mid-Sixties", *Journal of Farm Economics.* Vol. 47. No. 1. February 1965. pp. 1-16; and comments including the following from the *Journal of Farm Economics.* Vol. 47. No. 3. August 1965. pp. 843-857: William G. Brown. "An Appraisal of a Critical Appraisal"; William J. Donovan. "Current Status of Agricultural Economics as a Science"; Ernest W. Grove. "Basic Research in a Quasi Science"; Ronald L. Mighell. "Is Agricultural Economics a Science?"; Edward I. Reinsell. "Agricultural Economics is an Exact Science"; A. Allan Schmid. "Science, Art, and Agricultural Economics"; R. G. F. Sptize. "Agricultural Economics: Predictive or Prescriptive"; M. M Kelso. "The Author's Last Word — For Now".

60. Kiehl, Elmer R. "Integration of the Sciences for Effective Research", *Journal of Farm Economics,* Vol. XXXIX. No. 5. December 1957. pp. 1230-1231.

61. Knoblauch, H. C., E. M. Low and W. P. Myer. "State Agricultural Experiment Stations: A History of Research Policy and Procedure". USDA Misc. Pub. 904. May 1962.

62. Krupp, Sherman. "Analytic Economics and the Logic of External Effects", *American Economic Review.* Vol. LIII. No. 2. May 1963. pp. 220-226.

63. Kuznets, George M. "Theory and Quantitative Research", *Journal of Farm Economics.* Vol. 45. No. 5. December 1963. pp. 1393-1403.

64. Manderschied, Lester V. "Some Observations in Interpreting Measured Elasticities", *Journal of Farm Economics.* Vol. ILVI. No. 1. February 1964. pp. 128-136.

65. McKee, Dean E. "Our Obsolete Data Systems", *American Journal of Agricultural Economics.* Vol. 54. No. 5. December 1972. pp. 867-880.

66. McPherson, W. W. and J. E. Farris. " 'Price Mapping' of Optimum Changes in Enterprises", *Journal of Farm Economics.* Vol. XL. No. 4. November 1958. pp. 821-834.

67. McPherson, W. W. "Sources and Use of Technical Data in Farm Management Analysis", *Journal of Farm Economics.* Vol. XI. No. 4. December 1952. pp. 803 and 804.

68. Manley, William T. "An Essay on Federal-State Research Programs in Agricultural Economics: Need and Prospects for the Future in Agricultural Marketing", *Southern Journal of*

Agricultural Economics. Vol. I. No. 1. December 1969. pp. 119-122.

69. Motes, W. W. "Economic Research Programs for Community Development-Why Do It?", *Southern Journal of Agricultural Economics.* Vol. 4. No. 1. July 1972. pp. 9-13.

70. Nagel, Ernest. "Assumptions in Economic Theory", *American Economic Review.* Vol. LIII. No. 2. May 1963. pp. 211-219.

71. Paarlberg, Don. "Methodology for What?", *Journal of Farm Economics.* Vol. 45. No. 5. December 1963. pp. 1386-1392.

72. Paarlberg, Don. "Norman Borlaugh-Hunger Fighter". Foreign Economic Development Service, U.S. Department of Agriculture and U.S. Agency for International Development. PA 969. December 1969.

73. Parsons, Kenneth H. "The Logical Foundations of Economic Research", *Journal of Farm Economics.* Vol. XXXI. No. 4. Pt. 1. November 1949. pp. 656-686.

74. Peirce, J. R. "The Social Uses of Science", *American Scientist.* October 1954. pp. 495-497.

75. Plaxico, James S. "Agricultural Development Effort Assessment: The Colombian Case", *Southern Journal of Agricultural Economics.* Vol. 2. No. 1. December 1970. pp. 69-76.

76. Rehnberg, Rex and Garrison, L. C. Universidad Nacional-Instituto Colombiano Agropecuario, Medellin, Colombia.

77. Rose, Elbert M. "Selection of Problems for Research", *American Journal of Sociology.* Vol. 54. November 1948. pp. 219-229.

78. Schermerhorn, Richard W. "Identifying Extension's Marketing Clientele and Adapting Economic Information to Their Needs", *Southern Journal of Agricultural Economics,* Vol. 51. No. 1. December 1969. pp. 113-118.

79. Shaffer, J. D. "Some Conceptual Problems in Research On Market Regulations", in *Federal State and Local Laws and Regulations Affecting Marketing.* Regional Res. Bul. No. 168. North Dakota State University, September 1965.

80. Smith, Eldon, D. "University Assistance in Building Agricultural Research Institutions in Southeast Asia—A Case Analysis", *Southern Journal of Agricultural Economics.* Vol. 2. No. 1. December 1970. pp. 61-67.

81. Stern, Alfred. "Science and the Philosopher", *American Scientist.* July 1956.

82. Sundquist, W. B. "Federal-State Research Programs in the Economics of Agricultural Production — Needs and Prospects for the Future", *Southern Journal of Agricultural Economics.* Vol. 1. No. 1. December 1969. pp. 123-127.

83. Thirring, Hans. "The Steps From Knowledge to Wisdom", *American Scientist.* October 1956. pp. 445-446.
84. Vining, Rutledge. "Methodological Issues in Quantitative Economics: Variations Upon a Theme by F. H. Knight", *American Economic Review.* Vol. XL. No. 3. June 1960. pp. 267-284.
85. Weaver, Warren. "The Imperfections of Science". *American Scientist.* Vol. 49. No. 1. March 1961. pp. 99-113.
86. West, Quentin M. "The Changing Mission of Agricultural Economics Research". Division of Information, Economic Research Service, USDA. June 1973.
87. White, James H. "Relevant Research Areas and Organizational Questions Relative to Federal-State Research Programs in the Economics of Agricultural Production". *Southern Journal of Agricultural Economics.* Vol. 1. No. 1. December 1969. pp. 129-132.
88. Wilson, C. Peairs. "New Priorities in Agricultural Research — Implications for Economists in the West: An Administrative View", *Western Agricultural Economists Association.* Proceedings, 44th Annual Meeting. 1971. pp. 106.
89. Zimet, David Joseph. "The Economic Potential for Increasing Vegetable Production in the Zapotitan District, El Salvador". Unpublished M.S. thesis, University of Florida. 1974.

APPENDIX

AN APPLIED RESEARCH PROJECT PROPOSAL

I. Project Title: The Florida Shrimp Processing Industry: Economic Structure and Market Channels*

II. Project Leader: Jose Alvarez

III. Anticipated Duration: January 1, 1973 to December 31, 1973.

IV. Problem Statement:

    A.   Problematic Situation:

In 1970, Florida produced over $112 million of processed seafood products and was the leading state in the Southeast Region. Shrimp products, valued at $262 million, accounted for over 63 percent of the value of all processed seafood products in the Southeast Region in 1970. That year, Florida accounted for over 30 percent of all shrimp products processed in the Region.

Shrimp is the most important seafood processed in Florida. Total value of the shrimp processed in the State in 1970 was slightly over $79 million or over 70 percent of the value of non-industrial seafood products.

Florida's shrimp processing industry has been growing in both absolute and relative magnitudes during the last decade. The State processed about 34 million pounds of shrimp in 1960 which were valued at more than $25 million. More than 68 million pounds were processed at a value of $79 million in 1970. While Florida accounted for 19 percent of the total volume of processed shrimp consumed in the U.S. in 1960 and for 21 percent of value, both percentages changed to 25 by 1970.

Even though growth of Florida's shrimp processing industry during the last decade was substantial, shrimp landings in the state show a sharp decline from over 51 million pounds in 1960 to 29 million pounds in 1973. Only in 1960 and 1961 were landings greater than quantity processed. By 1970 Florida processed almost 96 million pounds and landed slightly over 31 million, representing a deficit of over 64 million pounds to be filled primarily by imports. Competition for imports, particularly with Japan and Canada, is becoming increasingly intense.

The importance of Florida in the U.S. shrimp processing industry, coupled with the growth of U.S. imports as a source of raw product for processors, suggests the importance of international shrimp trade both to U.S. consumers and Florida processors. Imports of

----

*This research project was completed with funding from Sea Grant. Special appreciation is extended to Dr. Jose Alvarez and Dr. Fred J. Prochaska, Food and Resource Economics Department, University of Florida, for their willingness to permit use of the proposal in this text and for their valuable assistance.

shrimp totaled 191.3 million pounds or 42 percent of total U.S. consumption in 1971. This was two million pounds less than the 193.3 million pound five-year average of 1966-1970.

B.   Problem:

A growing processing sector accompanied by a declining raw product sector causes the Florida shrimp processing industry to become increasingly dependent upon non-Florida shrimp for processing. The impact of past and potential changes in supply sources on market structure and ultimately on the competitive position of the Florida shrimp processing industry is uncertain and of concern to the industry and to the State because of the importance of the industry.

C.   Hypotheses:
1.   If the trend toward expanded imports of raw shrimp for processing continues, then the growth potential of the industry and the economic feasibility of locating shrimp processing firms in the State is reduced.
2.   Intensified competition for and changes in the source of raw product supply create structural changes in shrimp processing (size, number, and vertical and horizontal integration of firms) and changes in market coordination and performance (purchase and sale practices, channels and margins) which place pressure on small firms to exit from the industry. Related hypotheses are that:
   a.   As the age of firms in the industry increases, total sales increase thus providing a competitive advantage for these firms.
   b.   Large firms tend to be more involved than small firms in research, innovation and product promotion thus providing a competitive advantage for these firms.

D.   Objectives:
1.   Develop basic information to describe the present characteristics (size, employment, volume, value, etc.) of the firms in the Florida shrimp processing industry.
2.   Describe raw product supply, marketing channels and market structure in the Florida shrimp processing industry by studying the

81

product flow and principal components of the Florida shrimp procurement and marketing system.

3. Delineate the organization, market behavior, and market performance of firms in the Florida shrimp processing industry by studying entry and exit, market concentration and vertical integration.

4. Determine the extent to which firms are involved in research innovation and product promotion.

5. Identify emerging changes and trends in procurement, processing activities, and product markets which will influence the shrimp processing industry in Florida.

6. Provide information for firms within the shrimp processing industry about the feasibility for expansion given recent and expected changes in product procurement and market structure.

V. Procedures:

The market structure approach to research provides economists with a basic theoretical framework for studying the behavior of enterprises. Empirical research involving market structure according to W. F. Mueller[1], may be divided into three broad categories. The first is measurement of the various structural variables which economic theory suggests are relevant to the performance; second, empirical determination of the economic and technical bases of these variables; and third, specification and measurement of the interrelationships between market structure, firm conduct and industrial performance.

Secondary data will not be sufficient to satisfy all the objectives of this project so primary data will be collected in interviews obtained from Florida shrimp processors. An inventory of shrimp processing firms must be developed to determine the population to be interviewed.

A. Marketing Channels:

Specific information to be collected for industry description will include number and location of plants, type of ownership (if multi-unit or single-unit companies), organization of the procurement and marketing system, organization of production based on type of shrimp and other products processed, and plant size by plant

---

[1]Mueller, Willard F. "Empirical Measurement in Market Structure Research", *Journal of Farm Economics*, Vol. 43, December 1961, p. 371.

capacity, employment and 1972 production. The marketing system will be summarized by developing a product flow diagram of marketing channels which will describe the organization of the industry's supply, processing and distribution systems. These marketing channels will encompass the supply of fresh shrimp (Florida and other U.S.) and of frozen shrimp (U.S. and foreign), market activities carried on by the processing plants, and final market outlets.

Raw product procurement will be analyzed in terms of access of each firm to raw supply, the different purchase channels used, prices paid and the transportation employed to bring the raw supply to the plant. Processing activities and final product destination will be described.

B.   Evaluating Market Structure:

Market structure will be analyzed by studying the conditions of entry, market concentration, and market coordination. These major characteristics exert an influence on the nature of competition within the industry.

The condition of entry. Entry and exit activities during the 1961-71 period will be analyzed by using secondary data available from the Florida State Chamber of Commerce. Entry related to age of firms will be studied by testing the possibility of a relationship between total sales and the number of years in the industry. Finally, processor responses will provide the information for discussing present entry conditions.

Entry and exit activities over the 1961-71 period will be analyzed by developing a plant entry activity ratio, plant stability ratio and plant exit activity ratio. Exit related to age of firms will be tested using ordinary least squares methods. Processor responses will provide information relative to entry conditions.

Market Concentration. Market concentration will be determined with data obtained from each firm's statement of annual volume and value for each type of product processed. The procedures employed will be the Lorenz Curve and the Gini Coefficient to help analyze industry concentration ratios developed for the entire industry and each type of processed product.

Conditions of scarcity and control of Florida raw shrimp stimulate the need to study the number of sellers as buyers of raw products because this situation may affect the structure of the industry. To determine the nature of this potential relationship, the percentage of firms controlling Florida's raw shrimp will be compared with overall concentration by product types to account for some differences in product line among firms.

Market Coordination. Degree of vertical integration prevailing within the industry will be determined by the number of related

83

functions each firm performs. Functions may range from owning boats or unloading houses, operating breading plants, making boxes, owning transportation facilities and owning retail outlets to owning preprocessing facilities outside of the U.S.

Degree of horizontal integration will be established by determining the extent to which other seafoods were processed by the firms and the existence of processing facilities outside of Florida or the U.S. Reasons for horizontal integration will be explored.

C.   Evaluating Market Conduct:

Three measures of market conduct will be considered. These are pricing behavior, product strategy and product promotion.

Pricing behavior. Price determination by processors for raw shrimp in the fresh domestic market and in the frozen market will be studied. The possibility of price leadership will be explored and compared with concentration and backward integration into the ownership or control of raw supply.

Pricing methods for finished products by firms in the industry will be analyzed also. Pricing behavior will be compared with size of firms. Analysis of inventories will be related to price variations.

Further analysis will examine the relationship between price per unit and the number of units sold by each firm. Ordinary least square analysis will be used to investigate the hypothesis that price differentials prevail among firms in the Florida shrimp processing industry. The hypotheses to be tested will be that the price remains constant at all levels of quantity sold or that the price increases or decreases as quantity sold increases.

Product strategy and innovation. Product variation over time, in both quality and design, will be analyzed to determine product policy. Product variation will be examined in relation to size of firms and type of firms. Data will be collected on the effectiveness of large firms as compared to small in creating and introducing technological innovations. The chi-square statistical test will be used to determine if research and innovation are independent of firm size. Processor responses will be used to list product policy activities of the firms and to make comparisons with type of firms.

Product Promotion. Market structure theory suggests that fewness of sellers produces more incentives to advertise than large number of sellers. Since the Florida shrimp processing industry is composed of few sellers, this principle will be investigated using processor responses concerning advertising activities. Advertising will be compared with size of firms by means of the chi-square statistical test to determine if large firms are more likely to advertise than small firms in the industry.

D.   Some Measures of Market Performance:

Margin analysis. Margin analysis for the industry and for individual firms within the industry will be accomplished to determine market performance. Gross margins will be analyzed by using ordinary least squares methods to regress total revenue and total expenditures on the quantity of raw shrimp bought. The resulting equations then will be compared with the average revenue and expenditure functions for the industry in order to explain possible difference in margins. Margins also will be related to size of firms by means of the chi-square statistical test.

Productivity considerations. An attempt will be made to measure productivity as a means to evaluate performance. The percentage of plant capacity utilization during 1972 can be computed for the industry and for firms and for each group of firms with the same product lines will be computed.

E.   Possible Industry Changes:
Specific issues to be investigated in analyzing emerging changes and problems that may affect the industry include changes in raw product supply sources, changes in transportation and customers, problems with the supply of labor, utilization of plant capacity, and growth of firms over time.

VI. Budget:

Salaries	$6,220.00
Expendable equipment & supplies	150.00
Travel	1,000.00
Publication costs	150.00
Computer	200.00
Telephone	150.00
TOTAL	$7,870.00
Time available	One Year

VII. Published materials resulting from the project include:
Alvarez, Jose. "The Florida Shrimp Processing Industry: Economic Structure and Marketing Channels", unpublished M.S. Thesis, University of Florida, 1974.
Prochaska, Fred J. and Chris O. Andrew. "Shrimp Processing in the Southeast: Supply Problems and Structural Change," *Southern Journal of Agricultural Economics*, Vol. 6, No. 1, July 1974.
Alvarez, Jose, Chris O. Andrew and Fred J. Prochaska. "Dual Structural Equilibrium in the Florida Shrimp Processing Industry", *Fishery Bulletin*, National Oceanic and Atmospheric Administration, Vol. 74, No. 4, 1977 pp. 879-885.
Alvarez, Jose, Chris O. Andrew, and Fred J. Prochaska. "Economic   Structure of the Florida Shrimp Processing Industry", Florida Sea Grant Research Report, No. 9, 1975.

Andrew, Chris O., Fred J. Prochaska and Jose Alvarez. "Florida Shrimp: From the Sea Through the Market", Marine Advisory Program, Florida Cooperative Extension Service, SUSF-SG-75-005, May 1975.

INDEX

Analysis. 61
   complex 7
   consistancies 41
   methods 5
   nature 7
   regression 39, 40, 66
   sensitivity 67
   subjective 67
Analytical
   approach. 5
   proceedures. 5, 31, 61
   stages. 23
   techniques. 4, 23, 24, 38
   tools 55
Applied research . . . . x, 3, 80
Basic research 6, 46
Beliefs and values 50, 51, 60
"Brush fire" research. 12
Budget. 5, 31, 85
Calculators 11, 57
Case studies. 47
Castle, E. N. 17, 76
Change in definition 59
Code book. 58
Coding. 52, 58
Comparability. 41, 58, 59
Communication. 50-51, 60, 69
Computors 11, 57
Conceptual model. 14-16
Conclusions. 8, 34, 61
Confidence see levels of confidence
Confidentiality 57
Control 34, 42, 45, 47, 58
Criteria 63
Cross-section data 8, 9, 59
Cultural differences. .50, 51, 60

Data see also information
    acreage 9
    adaptation....................................... .41
    adjusted. .... .41
    analysis .... .34, 61
    combinations. .... .42
    comparability .... .41
    corrections .... .57
    cross section 8, 9, 59
    experimental................................ .8-10, 34-46
    limits. .... .57
    manipulation.................................... .57
    mix 8
    non-experimental 8-10, 47-60
    potential .... .57
    preservation.................................... .42
    primary...................... .8-10, 34, 47, 57-58
    processing 6, 48, 57
    production 9
    retrieval. 51-52
    review .... .57
    secondary 8-10, 34, 38, 41-42, 58-60
    selection .... .41
    time series..................................... .8, 9, 58
    transformation .... .57
    utilization .... .11, 61
Deadline 7, 61, 71
Debriefing. .... .53
Decision makers 3
Decision making process .... .70, 71
Decisions. 6, 7, 12, 61
Definitions .... .59
Degrees of confidence see levels of confidence
Demonstration trial................................ .42, 43
Errors .... .57
    interpretation .... .63
    reporting. .... .54
    response .... .52
    sampling .... .54
    Type I. 67-68
    Type II 67-68
Exploratory experiment .... .40
Estimates. .... .66
Experience 5, 10

Experimental
  control 42, 45
  data..8-10, 34-46
  design 35-38, 46
Factors 34
  independent 38
Field conditions 5
Financial
  resources.......................................12
  restrictions 6
Flexibility........................46, 49, 57, 60, 62, 68
  interpretation 63
Focus 14, 18
Heterogeneity 50, 53
Hildreth, R. J....................................17, 76
Homogeneity.......................................53
Human resources................................10-11
Hypotheses 5, 11, 14-16, 18, 67, 68
  characteristics 23
  formulation..............................4, 16, 22, 23
  null...66
  testable 17-19, 21, 23, 31
  testing.......................................34, 55
Hypothetical......................................21
Ideas..16
If-then
  propositions 23
  relationships 24, 31
Information see also data 5, 6, 14, 16, 18, 20, 28, 38, 43, 55
  primary..8
  relevant..2
  resources 7-10
  secondary 8
  useful 3
Interpretation 5, 50, 55, 61
  flexibility 63
Interrelated components...............................5
Interviewers................................11, 53, 56-57
Interviews 47, 53, 56
Language
  problems.......................................50
  non-scientific...................................70
Levels 34, 35, 47
  significance 66
Levels of confidence 6, 7, 12, 24, 40, 48, 49, 66, 68, 69
Levels of precision..................................6

Limits

    data..57

    definition 32

Linear program 10

Literature

    review 18

Long run measures 58

McPherson, W. W..............................66, 77

Manpower see personnel

Meaning..61, 63

    results 63

    data...69

Means 32

Measures 7, 34, 47

    measurement....................................9

    long run.......................................58

    short run......................................58

Methodology..4

Methods of analysis................................5

Model 10

    conceptual................................. 14-16

Modifications 8

Multidisciplinary

    experimentation 45

    cooperation....................................45

Multi-purpose experimentation......................42

Nature...34, 48

Needs 16, 17, 19, 21, 31, 71

Non-experimental data 8-10, 34, 47-60

Non-hypothetical 17-18, 31

Objectives.......... 4, 5, 14-16, 22, 32, 38, 47, 55, 56, 70, 81

    delineation 28-29, 71

Objectivity 46, 56

Observations 7, 38

Orientation 22

    projects.......................................14

Personnel......................................5, 11

    trained..11

    manpower......................................6

Physical resources 6, 11

Plan of execution 15, 16

Planning 5, 6, 8, 10, 11, 45

Population. 48
  survey 53
  heterogeneity 50, 53
  homogeneity 53
  target. 48
Practical experience. 5
Practice. 67
Precision 6, 7, 9, 40, 46
Precoding 52
Pretesting 52-55, 60
Primary data 8, 34, 47, 57-58
Problem. 81
  definition 4, 19
  identification. 4, 7
  manageable 19, 77
  orientation 3, 35
  non-hypothetical. 17-18
  relationship to design 85
  relevant. 17, 19
  researchable. 14-20, 31
  resolution 4, 23, 29, 69
Problem statement 5, 18-23, 31, 81
Problematic situation 15, 17, 19, 20, 31, 80
Procedures. 5, 14, 28, 29, 31, 61, 82
Product
  useful 16, 71, 72
Project
  focus. 14
  orientation 43
  proposal 80
  scope. 43
  size 43
Project statement 14, 17, 81
Proposal 30-31, 80
Publication 31, 55, 85
  forms. 70
Questions
  grouping 52
  open-ended 51
  organizing 52
  sensitive. 52
Questionnaires. 47, 51, 60
  design 49-50, 60
  mail. 55
  pretesting 52-55, 60

Rapport. 50
Recommendations. 34, 61, 69, 71
Regression analysis 39, 40, 66
Relationships. 16, 18, 30, 63
   nature 41
   problem. 23, 35-38
   resources 38-41
Relevance 8, 31, 49
Reliability 63, 66-69
Replications 7, 35, 40, 45
Reports 5, 69, 71
   technical 70
Resolving problems 4, 6, 13, 16, 23, 28, 32, 34, 69
Respondents 50-51, 53
   selection 47-48, 56
Responses 51-56
   critical review 55
   errors. 52
   modifications 55
   selective. 52
   semi-quantitative. 51
Resource availability 6-13, 15, 29, 41, 45, 72
   financial 12
   human. 10-11
   information. 7-10
   physical. 11-12
Resource restraints 3-13, 14, 31, 45, 46, 47, 50
   budgetary 5
   financial 6
   information. 6
   institutional. 5
   manpower. 6
   physical facilities. 6
   time, ... 6, 10, 12-13, 38, 41, 47, 50, 55, 72
Resources
   relationship 38-40
   research. 3, 24
Results 5, 23, 24, 31
   form 69-71
   meaning. 63
   presentation 69, 71
   reliability. 66, 63
   understanding 62, 69
   useful 13, 16

Review 55, 69, 71
Risk. 66
Sample 7, 47, 60
  design 53
  heterogeniety 50
  random 48
  sequential 49
  stratified 48
Secondary experimental data 34, 38, 41-42, 58-60
Sequential sampling technique 49
Series
  price 8, 9
  production 8
  time. 8, 9
  verifying 58
Short run measures 58
Significance levels 66
Simulation model 10
Social organization 48
Statistical
  analysis 11
  precision 38
Statistician 38, 47
Stratification. 47
Subjective
  analysis 62
  evaluation 47
Supervision 43
Survey
  informal 10
  sample. 47
Tabulation. 55
Target population 48
Techniques 23, 24, 47
  analytical. 38
  interviewing. 56
Telescoping bias 60
Terminology 50, 60
Theoretical approach. 5
Theory 4, 23, 24
Theses. 12, 70
Time constraints 6, 10, 12-13, 38, 41, 47, 50, 72
Time series 8, 9, 58
Time sequences 5, 31

Training. 11, 47, 56-57
Translation 50
Treatments 35, 40, 43, 45
Validity. 18
Variance 10, 49, 53
    analysis 39, 40, 66
Variables. 7, 9, 24, 34, 47
    aggregate 7
Verification. 24, 51, 57, 58
Work plan 5
Work schedule. 55